# CADMIUM TELLURIDE QUANTUM DOTS

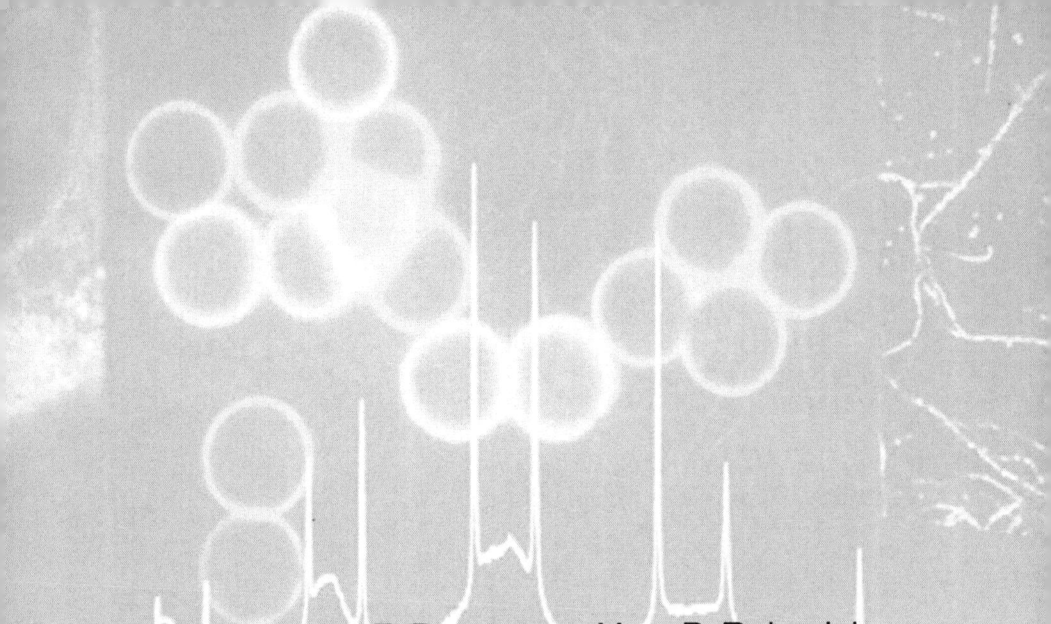

edited by John F. Donegan Yury P. Rakovich

# CADMIUM TELLURIDE QUANTUM DOTS

## ADVANCES AND APPLICATIONS

PAN STANFORD PUBLISHING

*Published by*

Pan Stanford Publishing Pte. Ltd.
Penthouse Level, Suntec Tower 3
8 Temasek Boulevard
Singapore 038988

Email: editorial@panstanford.com
Web: www.panstanford.com

**British Library Cataloguing-in-Publication Data**
A catalogue record for this book is available from the British Library.

**Cadmium Telluride Quantum Dots: Advances and Applications**



ISBN 978-981-4316-05-7 (Hardcover)
ISBN 978-981-4316-09-5 (eBook)

Printed in the USA

# Contents

# Foreword

The field of semiconductor nanocrystals (colloidal quantum dots) is booming nowadays, and there are plenty of books in the area. At the same time, there are not too many quantum dot systems which would attract sufficient attention to allow for a specific book covering a specific material. CdTe quantum dots, which are in focus of this book, are one of these rare cases. The authors of this book offer their profound knowledge related to the synthesis, assembly, optical properties and biological applications of mainly thiol-capped, water-based CdTe quantum dots. I am particularly proud of this book as it illustrates the tremendous development in the field which I was fortunate to start with our very first publication on this particular quantum dot system, which appeared in the German journal *Berichte der Bunsengesellschaft für Physikalische Chemie* in 1996. Since then, scientists worldwide have achieved great progress in understanding and refining many aspects related to CdTe nanocrystals, and the editors of this book have done a great job for the readers who are (and will be) interested in this particular quantum dot system.

Enjoy!

**Andrey L. Rogach**
City University of Hong Kong
Hong Kong

# Preface

Semiconductors are perhaps the most important of all the advanced materials that we use in our everyday lives. Silicon is the archetypal material that is now processed to an extraordinary degree into the huge number of transistors that we all have in our home computers, laptops, and iPhones. Other semiconductors such an indium phosphide are used in lasers that are employed in optical communication systems that power the Internet. The II-VI semiconductors are well known as light-emitting structures, but their applications are not so well developed. However, the research in these materials is very vibrant, and a large number of publications are related to potential applications in solar cells, biosensing, and light-emitting diodes. Of the II-VI materials, cadmium telluride (CdTe) has received a large amount of attention, particularly in materials that are fabricated with colloidal synthetic methods.

Quantum dots or nanocrystals are nanoscale semiconductor materials in which the electronic properties are strongly changed due to quantum confinement effects. These quantum dots exhibit the well-known effect of having a bandgap energy that depends on the size of the dot itself. Cadmium telluride quantum dots fabricated by colloidal synthesis are the subject of our book. In this book, we present contribution by some of the leading researchers throughout the world in this field.

The book starts with a chapter by Núñez Fernández and Vasilevskiy on the electronic and optical properties of quantum confined cadmium telluride materials, followed by two chapters on synthesis methods for colloidal quantum dots. In the first of these chapters (Lesnyak, Gaponik, and Eychmüller), a detailed study of the synthetic methods to produce highly luminescent CdTe quantum dots is provided. In the second, Gaponik describes various methods to form assemblies of CdTe quantum dot structures. Energy transfer between quantum dots and between dots and other nanoscale systems is an area of very high research activity at present. In Chapter 4, Lunz and Bradley give a detailed explanation of the Förster energy transfer process and how this is studied in quantum dot systems. In Chapter 5, Rakovich and Donegan describe in detail the changes that occur to the emission properties of CdTe quantum when they are coupled to microcavity structures in which whispering gallery modes are observed. Chapters 6 and 7 describe the applications of CdTe quantum dots as sensors in biological systems. LeCigne, Sukhanova, and Nabiev explain how energy transfer processes are altered by coupling CdTe quantum dots to particular biological systems, including bacteriorhodopsin. Williams, Conroy, and Volkov deal with a key subject: the biocompatibility of cadmium telluride when placed in close contact with living cells.

This book will be of great interest to a wide range of physicists, chemists, engineers, biologists, and materials scientists working in the area of science and applications of cadmium telluride quantum dots. In particular, those who work at the interface between the various disciplines mentioned above will find the book very helpful. In addition, graduate students who are working in these fields will find the book very easy to follow, and the large number of references will assist in building their understanding of these fields of research. We would like to thank the authors for their excellent contributions and also Andrey Rogach (City University of Hong Kong) for writing the foreword to the book. Finally, we thank Stanford Chong for his assistance as we worked on getting this project through to completion.

**Yury P. Rakovich**
Donostia—San Sebastian

**John F. Donegan**
Dublin

Autumn 2013

Chapter 1

# Optical Properties of Bulk and Nanocrystalline Cadmium Telluride

**Yuriel Núñez Fernández and Mikhail I. Vasilevskiy**

*Centre of Physics and Department of Physics, Minho University, Campus de Gualtar, Braga, P-4710-057, Portugal*

mikhail@fisica.uminho.pt

## 1.1 Bulk Crystals

Since the 1960s, cadmium telluride has been extensively studied for its potential as a nuclear radiation detector, an electrooptical modulator, a solar cell, and an optical material in the infrared. The results of the earlier work have been reviewed in the book by Zanio [1]. Since then, some new data have been published, concerning, in particular, the calculated electron and phonon spectra and measured exciton properties. Here we give a brief overview of the material, electronic and optical properties of bulk CdTe, relevant to its properties in the nanocrystalline form.

Under normal conditions, cadmium telluride crystallizes in the zinc blende (cubic) structure [1–3], while the wurtzite (hexagonal) structure has also been observed in thin films [4]. At

*Cadmium Telluride Quantum Dots: Advances and Applications*
Edited by John F. Donegan and Yury P. Rakovich

ISBN 978-981-4316-05-7 (Hardcover), 978-981-4316-09-5 (eBook)
www.panstanford.com

high pressure (near 3.5 GPa), it undergoes two closely spaced transitions: first zinc blende to cinnabar and then cinnabar to NaCl [5]. Nevertheless, the zinc blende structure (symmetry group $T_d$; see Fig. 1.1) is the standard one for CdTe crystals. The reciprocal lattice is bcc and the first Brillouin zone (BZ) is shown in Fig. 1.2.

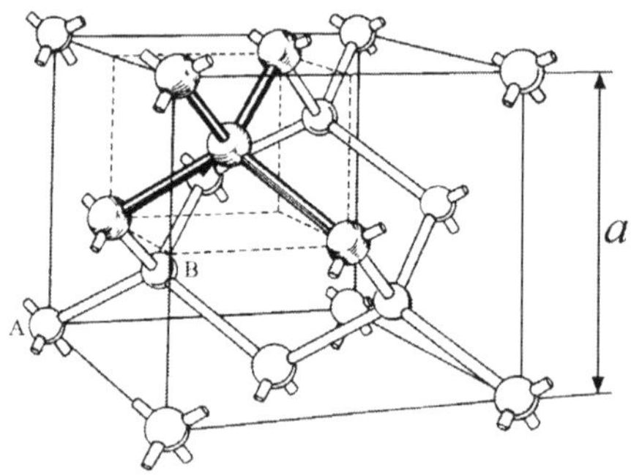


**Figure 1.1** Zinc blende crystal structure of CdTe. Cations (Cd) and anions (Te) occupy two face centered cubic sublattices (A and B). Lattice parameter $a$ = 0.6481 nm ($T$ = 300 K), linear expansion coefficient (4.8 ± 0.1) × $10^{-6}$ $K^{-1}$ [1–3].

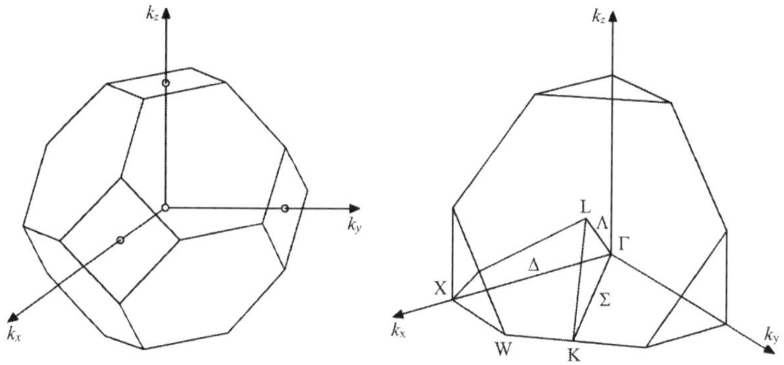


**Figure 1.2** First BZ for zinc blende crystal structure and its irreducible part showing high symmetry points and directions in the **k**-space.

The electronic band structure of CdTe has been calculated using several numerical approaches, such as empirical [6] and *ab initio* pseudopotential methods [7,8] within the framework of density functional theory (DFT) and the tight-binding approximation [9]. Figure 1.3a shows this structure, typical of cubic II–VI and III–V semiconductor compounds. Several recent calculations

[10,11] performed using the Vienna *Ab initio* Simulation Package (VASP) confirm the earlier results. In brief, CdTe is a direct gap semiconductor with the smallest energy gap located at the center of the Brillouin zone (Γ). The conduction band ($\Gamma_6$) is formed by *s*-type Cd orbitals. It is nearly parabolic (with the effective mass $m_e \approx$ const) for energies up to 1 eV above the bottom ($E_{\Gamma_6}$) [10]. The topmost valence band exhibits *p*-like orbital symmetry and is split by the spin–orbit coupling into a fourfold ($\Gamma_8$) and a twofold ($\Gamma_7$) state. The former, corresponding to the total angular momentum $J = 3/2$ and degenerate in the Γ point, gives rise to light hole and heavy hole sub-bands (with effective masses $m_{hh} > m_{lh}$ ) for $\mathbf{k} \neq 0$ (Fig. 1.3b).

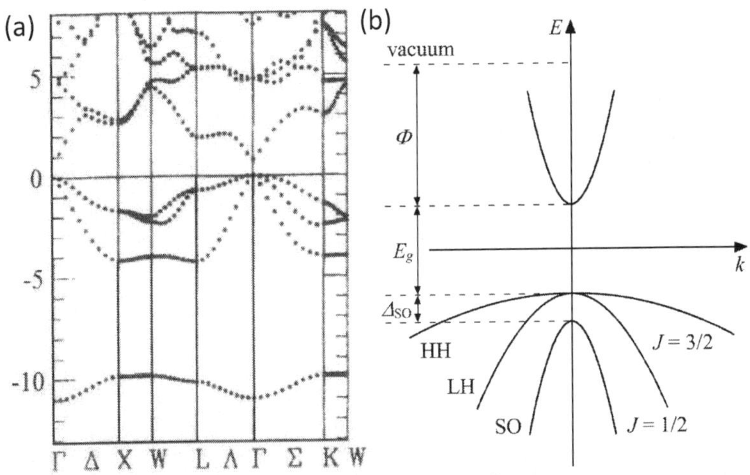


**Figure 1.3** (a) Band structure of CdTe calculated using an *ab initio* pseudo-potential method [8] (reproduced with permission of the American Physical Society). (b) Essential features of the valence and conduction bands near the Γ point. The electron affinity, Φ, is 4.28 eV [1].

Experimentally, the optical properties determined by the electronic band structure (UV-visible spectral region, typically 5.5 to 1.5 eV) have been measured using several methods, such as spectral ellipsometry [12], reflectivity [13], angle-resolved photoemission [14], and Faraday rotation [15]. For the bandgap energy, $E_g = E_{\Gamma_6} - E_{\Gamma_8}$, at room temperature (RT) values ranging from 1.49 to 1.53 eV have been reported [16]. It seems that the actual generally accepted RT value of $E_g$ is 1.512 ± 0.003 eV [17]. The value at liquid-helium temperature is well established as 1.606 eV

[1,2,17]. According to Ref. [17], the temperature dependence of the bandgap energy in the range 4–500 K is best fitted by the expression proposed by Pässler in Ref. [18], although the most popular empirical relation proposed by Varshni [19],

$$E_g(T) = E_g(0) - \frac{\alpha_g T^2}{(\theta + T)}, \tag{1.1}$$

still can be used, with the parameters $\alpha_g$ = 0.372 meV/K and $\theta$ = 110 K. A value of (6.5 ± 0.2) × $10^{-6}$ eV·$bar^{-1}$ has been obtained for the pressure coefficient, $dE_g/dp$ [9].

The spin–orbit splitting energy is $\Delta_{SO} \approx 0.95$ eV [14]. Higher energy features in the spectrum of the complex dielectric function, corresponding to the van Hove critical points lie at $E_1 \approx 3.3$ eV, $E_1 + \Delta_{SO} \approx 4$ eV and $E_2 \approx 5.5$ eV [13]. Some further band structure parameters (average of the values that were either calculated or measured experimentally, taken mostly from Refs. [1,2]) are listed in Table 1.1, where $P$ denotes the dipole moment matrix element between the $\Gamma_8$ and $\Gamma_6$ bands. This inter-band transition is dipole-allowed and determines the optical absorption of CdTe in the visible range.

**Table 1.1** Some band structure parameters at $T$ = 300 K

| Parameter | $E_g$ (eV) | $m_e$ | $m_{lh}$ | $m_{hh}$ | $P$ (eV·cm) |
|---|---|---|---|---|---|
| Value | 1.512 | 0.10 | 0.12 | 0.5 | 7 × $10^{-8}$ |

The exciton effect near $E_g$ is weaker than in other II–VI semiconductors (such as CdS or ZnO) but still quite significant. The exciton Bohr radius is $a_{ex} \approx 7.3$ nm [20] and the binding energy is $R_{ex} \approx 10.1$ meV [17]. Free exciton photoluminescence (PL) line and its phonon replicas have been observed [16,17,21] up to room temperature. Moreover, it was shown that even at higher temperatures, $T$ > 300 K, discrete excitonic transitions still strongly influence the absorption edge [18] and its shape deviates from the simple dependence,

$$\alpha(\omega) \propto \frac{\sqrt{\hbar\omega - E_g}}{\omega}, \qquad (\hbar\omega \geq E_g) \tag{1.2}$$

expected for direct transitions within the non-interacting electron–hole model [22]. It was fitted using a combination of exponential and Lorentzian functions [17] (resembling the Elliott formula valid for excitonic absorption at zero temperature [22]).

Below the fundamental absorption edge ($\hbar\omega < E_g$), the intrinsic material is essentially transparent down to the region of the optical phonon response in the far-infrared (the longitudinal optical phonon energy $\hbar\omega_{LO} \approx 21$ meV [2]). The refractive index, $\eta$, was measured in the region 1.45–0.5 eV and fitted using the formula [19]

$$\eta^2 = A + \frac{B\lambda^2}{(\lambda^2 - C^2)}, \tag{1.3}$$

with $A$ = 5.68, $B$ = 1.53, $C^2$ = 0.366 μm² and the wavelength measured in μm. Notice that at large $\lambda$, $\eta^2 \to A + B = \varepsilon_\infty$, the high-frequency dielectric constant (see Table 1.2).

In the spectral region corresponding to the optical phonon response, the dielectric function is well represented by the usual Lorentzian dispersion function,

$$\hat{\varepsilon}(\omega) = \varepsilon_\infty + \frac{4\pi\rho}{\omega_{TO}^2 - \omega^2 - i\Gamma_{TO}\omega}, \tag{1.4}$$

with the oscillator strength $\rho = (\varepsilon_0 - \varepsilon_\infty)\omega_{TO}^2/4\pi$, where $\varepsilon_0$ is the static dielectric constant. The transverse (TO) and longitudinal (LO) optical phonon frequencies (in the Γ point) are rather well established using the far-infrared (FIR) reflectivity and Raman spectroscopy techniques [2,22]. They obey the Lyddane–Sachs–Teller relation, $\omega_{LO}^2/\omega_{TO}^2 = \varepsilon_0/\varepsilon_\infty$. The values of these parameters are summarized in Table 1.2 where the transverse and longitudinal sound velocities [23] are also presented. The phonon dispersion curves have been calculated [24,25] as well as measured using the neutron scattering method [26]. As typical for semiconductors with zinc-blend structure, the LO phonon branch is bended downwards (with the X point frequency ≈ 135 cm⁻¹), while the TO phonon branches are almost flat.

**Table 1.2** Phonon parameters and dielectric constants at $T$ = 300 K

| Parameter | $v_T$ (km·s$^{-1}$) | $v_L$ (km·s$^{-1}$) | $\omega_{TO}$ (cm$^{-1}$) | $\omega_{LO}$ (cm$^{-1}$) | $\varepsilon_\infty$ | $\varepsilon_0$ |
|---|---|---|---|---|---|---|
| Value | 1.7 | 3.4 | 140 ± 0.5 | 168 ± 0.5 | 7.2 | 10.3 |

## 1.2 CdTe Nanocrystal Quantum Dots

### 1.2.1 General Remarks

CdTe nanocrystals (NCs) possessing the properties of quantum dots have been produced using a number of techniques, such as growth in special glasses by quenching melts [20,27], and chemical synthesis in colloidal solutions, either organic [28,29] or aqueous [30]. While the melting approach yields just nearly spherical particles, the colloidal chemistry techniques can be used also to fabricate very small clusters of some "magic" sizes (for example, $Cd_{54}$ $Te_{32}$ $(SCH_2CH_2OH)_{32}^{8-}$ particles [31]), nanorod and tetrapod morphologies [32], core/shell structures such as CdTe/CdS and alloyed materials such as CdSeTe and CdHgTe in the nanocrystalline form (see Chapter 2). As with nanocrystal QDs of other II–VI materials (CdS and CdSe), they show size-dependent exciton transitions in the absorption and PL spectra. This dependence is discussed in some detail in the next section. The chemically grown NCs, capped with a protecting organic shell, keep their high PL quantum yield when incorporated into a polymer (e.g., PMMA) matrix or cast onto a substrate in the form of a matrix-free film (see Fig. 1.4). As usual, the PL peak is normally Stokes-shifted with respect to the lowest energy absorption peak (however, under special excitation conditions, a PL up-conversion can take place [33–35]; see Section 1.2.4). Even though the fine structure of the QD exciton state can produce some Stokes shift for the single dot emission, the main reason for it in QD ensembles is the size dispersion effect [36].

The non-linear properties of CdTe NCs have been studied in Refs. [37,38] and quite a high value of the non-linear refractive index, $n_2 = -6 \times 10^{-13}$ cm$^2$W$^{-1}$, has been measured [38], which is higher than for all other NC materials studied in that work. The phonon-related optical properties (i.e., Raman and FIR absorption spectra) are discussed in Section 1.2.3. Finally, Section 1.2.4 briefly describes the up-converted (or anti-Stokes) PL mentioned above, a phenomenon involving both excitons and phonons confined in a QD and also cascade excitation of several dots within a NC ensemble [39]. Note that another collective effect taking place in NC ensembles, the Förster-type resonant energy transfer without charge transport, is discussed in detail in Chapter 4.

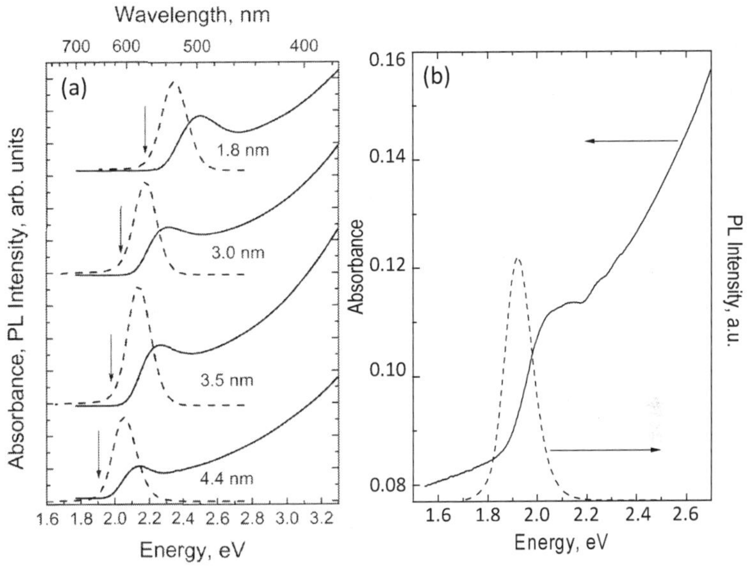


**Figure 1.4** RT absorption (full lines) and non-resonantly excited PL (dashed lines) spectra of chemically grown CdTe QDs with different average sizes dispersed in solution (a) and spectra of PMMA films containing CdTe QDs of ≈5.2 nm in size [33]. The arrows in panel (b) indicate the excitation wavelength in each case, yielding up-converted PL [34].

## 1.2.2 Size-Dependent Exciton Response

Since the exciton Bohr radius for CdTe is rather large (7.3 nm), in most cases the NC exciton states correspond to the strong confinement regime ($R << a_{ex}$). Then, in the first approximation, one can consider the spatial quantization of the energy spectra of non-interacting electrons and holes and the Coulomb interaction can be treated as a perturbation. A simple description of electron–hole pair states confined in a CdTe spherical QD is possible within the effective mass approximation (EMA), since the underlying bulk material possesses a cubic crystal lattice and a direct bandgap [40]. The detailed theory taking into account the complex structure of the valence band of bulk CdTe can be found in several original works [36,41] and recent reviews [39,42]. Assuming infinitely high barriers for both electrons and holes at the NC/matrix interface and a very large spin–orbit splitting, $\Delta_{\rm SO} \to \infty$, the energy of the exciton ground state, denoted by $(1s_{\rm e}\, 1S_{3/2})$ and 8-fold degenerate because of the electron spin and hole angular momentum ($J = 3/2$), is given by

$$E_{\rm ex}(R) = E_{\rm g} + \frac{\hbar^2\pi^2}{2m_{\rm e}R^2} + \frac{\hbar^2\chi_1^2}{2m_{\rm hh}R^2} - \zeta\frac{e^2}{\varepsilon_0 R}. \tag{1.5}$$

Here $\chi_1$ is a root of a certain characteristic equation that contains the only parameter $\beta = m_{\rm lh}/m_{\rm hh}$ [27,36]. For the effective masses of Table 1.2, $\chi_1 \approx 4.8$.

The last term in Eq. (1.5) represents the Coulomb energy and $\zeta$ is a coefficient of the order of unity. According to [43], the polarization effect produced by the electron–hole pair on the dielectric matrix surrounding the dot can also be included in this term. Then the coefficient $\zeta$ depends on the static dielectric constant of the matrix, $\varepsilon_{\rm M}$, according to the relation $\zeta = [(0.86\varepsilon_0 + 2.72\varepsilon_{\rm M})/(\varepsilon_0 + \varepsilon_{\rm M})]$ [43]. Note that the Coulomb attraction just shifts the energy of the lowest excitonic state, without lifting its degeneracy. Beyond Eq. (1.5), the electron–hole exchange interaction, within the EMA, scales as $R^{-3}$ and leads to a splitting of the $(1s_{\rm e}\, 1S_{3/2})$ octet into two groups of states, characterized by the total (exciton) angular momentum $F$ = 1, 2. The splitting can reach some 20–30 meV in small CdTe QDs [36,42].

It is known that the EMA overestimates the electron and hole confinement energies [44] and the scaling laws of these energies with the QD radius obtained by fitting experimental data do not precisely agree with the EMA prediction (1.5). The agreement can be improved if finite-height interface barriers [45,46] and the $\mathbf{k}\cdot\mathbf{p}$ interaction between the $\Gamma_8$, $\Gamma_7$, and $\Gamma_6$ bands [47] are taken into account. The former is particularly important when colloidal QDs capped with an organic shell (such as trioctylphosphine oxide, TOPO [28] or thioglycolic acid, TGA [30]) are considered. It has been possible to fit rather well the experimentally measured $E_{\rm ex}(R)$ dependence for CdTe QDs assuming low barriers ($\approx$1.6 eV) at the NC/TOPO interface [45,46], even though an oversimplified approximation for the valence band was used in these works.

Taking into account the light and heavy hole subbands and using generalized boundary conditions (allowing for the envelope wavefunctions to be discontinuous at the NC/cap shell interface), it is possible to achieve an excellent agreement with the experimentally measured optical gaps for NCs with $R$ as low as 1.5 nm [48] (see Fig. 1.5).

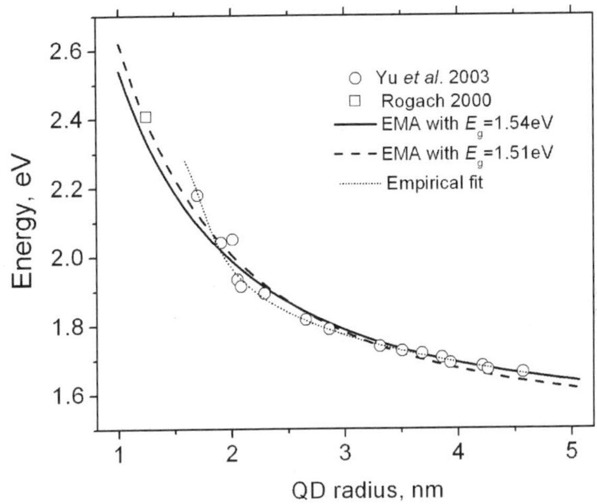


**Figure 1.5** Measured (points, Refs. [49] and [65]) and calculated (full and dashed lines for two values of bulk $E_g$) energies of the lowest exciton state for TOPO covered CdTe NCs. The dotted line is an empirical fit corresponding to Eq. (1.6).

Of course, the EMA fails in the ultimate limit of very small clusters containing a hundred of atoms that should not even be called nanocrystals because their properties have more similarity with molecules than with crystals. It is difficult to indicate the exact limit of validity of the EMA results. A systematic study of the dependence of the optical gap (absorption peak position) on the NC size (independently determined from transmission electron microscopy studies) was performed in [49,50] and the following empirical relation has been proposed for CdTe dots [49]:

$$2R[\text{nm}] = 9.8127\times10^{-7}\lambda_{\text{ex}}^{3} - 1.7147\times10^{-3}\lambda_{\text{ex}}^{2} + 1.0064\,\lambda_{\text{ex}} - 194.84, \quad (1.6)$$

where $\lambda_{ex} = 1.24 \times 10^3 / E_{ex}\,[\text{eV}]$ is the wavelength (in nm) corresponding to the lowest energy exciton transition. Although this formula has no physical grounds, it seems to work well in the size range of 1.5 to 5.5 nm [49] and is quite popular among users of CdTe NCs. However, it is necessary to bear in mind that the scaling law, $E_{ex}(R)$, depends on the type of the capping organic shell used [30,50], and the so-called "sizing curve" (like the one presented for CdTe in Ref. [30]) is not completely universal for a given NC material (this is the finite barrier height effect, in terms of EMA). In a recent work [51], not just the optical gap but also the size dependences of the highest occupied and lowest unoccupied electronic states in small size CdTe QDs stabilized by oleic acid were studied by means of cyclic voltammetry. For the smallest sizes ("magic size" CdTe clusters), the measured dependences are weaker than those predicted by the EMA. The gap widths tend to approach the values obtained by DFT calculations performed for nonstoichiometric $Cd_mTe_n$ ($m \neq n$) clusters [51].

The emission and absorption of light by a single QD (of radius $R$) is described by its susceptibility that takes into account the exciton transitions:

$$\chi(\omega,R) = \frac{4|d_{cv}|^2}{3V} \sum_n \left\{ \frac{|C_n(R)|^2}{E_n(R) - \hbar\omega - i\Gamma_n} + \frac{|C_n(R)|^2}{E_n(R) + \hbar\omega - i\Gamma_n} \right\}, \quad (1.7)$$

where the sum runs over confined exciton states with energies $E_n$ ($E_0 = 0$ denotes exciton vacuum), $\Gamma_n$ is the homogeneous broadening owing to the finite lifetime of the exciton state, $d_{cv}$ is the transition dipole moment matrix element between the valence and conduction bands of the underlying bulk material, $V$ is the QD volume and $C_n = g_n \int \Psi_{ex}^{(n)}(\mathrm{r}, \mathrm{r})dr$ with $g_n$ and $\Psi_{ex}^{(n)}(\mathrm{r}_e, \mathrm{r}_h)$ denoting the degeneracy factor and the wavefunction of the corresponding exciton states. In the first approximation, the latter is just a product of the envelope functions of the confined electron and hole.

The quantity $\varepsilon_{QD}(\omega,R) = \varepsilon_\infty + 4\pi\chi(\omega,R)$ can be regarded as a QD dielectric function, where $\varepsilon_\infty$ is the high-frequency dielectric constant of the QD material. One can also define the QD polarizability (assuming spherical shape of the dot),

$$A(\omega,R) = \frac{\varepsilon_{QD}(\omega,R) - \varepsilon_h}{\varepsilon_{QD}(\omega,R) + 2\varepsilon_h} R^3, \quad (1.8)$$

where $\varepsilon_h$ is the dielectric constant of the host material. The imaginary part of Eq. (1.8) determines the extinction cross-section and the emission properties of the dot. If the QD size and the typical inter-dot distance are much smaller than the light wavelength, one can calculate an effective dielectric function, $\varepsilon^*(\omega)$, of the composite material containing a certain volume fraction of QDs ($f$) embedded in the host matrix, using one of the following approaches: (i) Maxwell–Garnett approximation (MGA), valid in the low $f$ limit, (ii) Bruggemann mean field approximation (BA) valid when the QD volume fraction is not too small ($f \sim 0.5$). Both schemes can be generalized in order to include the QD size dispersion and yield closed relations for $\varepsilon^*(\omega)$,

$$\frac{\varepsilon^*(\omega)-\varepsilon_h}{\varepsilon^*(\omega)+2\varepsilon_h} = \frac{4\pi}{3} N \int A(\omega,R)F(R)dR, \tag{1.9}$$

and

$$f\int \frac{\varepsilon_{QD}(\omega,R)-\varepsilon^*(\omega)}{\varepsilon_{QD}(\omega,R)+2\varepsilon^*(\omega)} F(R)dR + (1-f)\frac{\varepsilon^*(\omega)-\varepsilon_h}{\varepsilon^*(\omega)+2\varepsilon_h} = 0, \tag{1.10}$$

in the MGA and BA, respectively. Here $F(R)$ is the QD radius distribution function (usually taken as Gaussian), $\varepsilon_h$ is the host matrix dielectric constant and $N$ is the total number of QDs per unit volume. MGA can be extended up to $f \sim 0.1$ using a renormalized polarizability that takes into account the dipole-dipole interactions between the dots polarized by the electromagnetic wave [52]. Once the composite dielectric function is known, the spectra of the sample containing QDs can be calculated using the standard optics formulae. For instance, for a sufficiently thick slab containing QDs (so that the interference effects are unimportant), the absorption coefficient is simply

$$\alpha(\omega) = \frac{2\omega}{c}\operatorname{Im}\sqrt{\varepsilon^*(\omega)}, \tag{1.11}$$

where $c$ is the velocity of light. Figure 1.6 shows the absorption spectrum of a borosilicate glass slab containing $f \approx 0.01$ (fitted value) of CdTe NCs of mean size 2.8 nm [27]. The dielectric function $\varepsilon^*(\omega)$ was calculated using the renormalized MGA scheme and assuming the Lifshitz–Slyozov size distribution function.

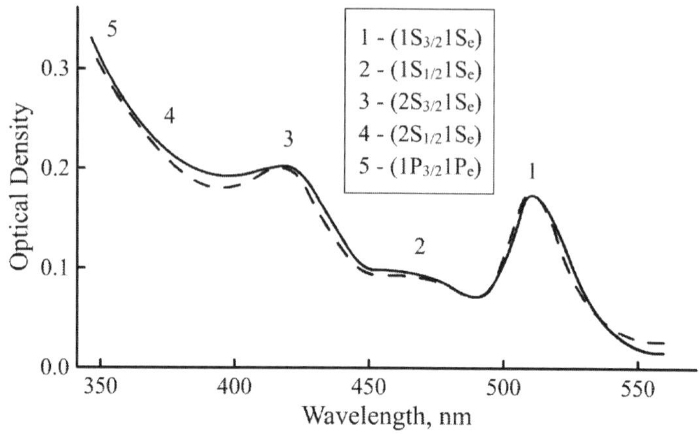


**Figure 1.6** Measured (full line) and calculated (dashed line) absorption spectra of CdTe NCs embedded in borosilicate glass. Several size-quantized transitions (indicated on the plot) were identified and included in the calculation [27].

### 1.2.3 Phonon Response: FIR and Raman Spectra

The general properties of phonons confined in nanostructures, including QDs, have been described in several reviews [39,53,54]. In particular, low-frequency acoustic phonon modes in a spherical nanoparticle are not really spatially quantized, even though the local (QD) density of these modes shows oscillations originating from the interference phenomenon between the outgoing acoustic wave and the one reflected at the QD/matrix interface [39]. Since this reflection is only partial, no complete confinement occurs. However, the oscillations of the QD phonon density of states with the wavenumber can be probed by the low frequency Raman scattering and some other techniques. This effect has been observed for CdTe NCs [55] demonstrating that the lattice dynamics of these small particles can be rather well described using the elastic constants (or, equivalently, the sound velocities) of the bulk material.

Turning to the optical phonons, these are supposed to be spatially quantized, similar to excitons. This has been demonstrated theoretically using a number of continuum models overviewed in [53,54]. The confined optical phonon modes in a spherical QD can be classified according to the spherical quantum numbers.

However, owing to the relatively small dispersion of the optical phonon branches of bulk CdTe, it is not easy to spectrally resolve the different quantized phonon modes. Figure 1.7a presents a FIR transmission spectrum of a matrix-free film of chemically synthesized CdTe QDs deposited on a silicon substrate [56]. The absorption band centered at 156 cm$^{-1}$ is related to the so-called Fröhlich mode, with the frequency determined by the LO and TO phonon frequencies of the semiconductor and the dielectric constants of the QD and matrix materials,

$$\omega_F = \sqrt{\frac{\varepsilon_\infty \omega_{LO}^2 + 2\varepsilon_h \omega_{TO}^2}{\varepsilon_\infty \omega_{LO}^2 + 2\varepsilon_h}}, \tag{1.12}$$

This absorption band is a consequence of the dielectric polarization effects at the nanoparticle/matrix interface, with no quantum size effects present [57].

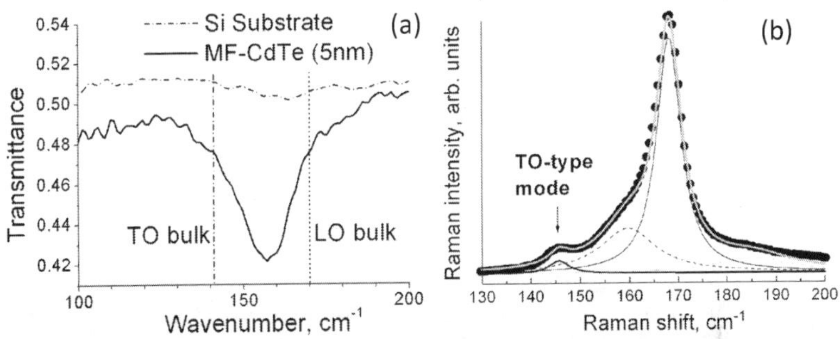


**Figure 1.7** (a) Experimental FIR transmittance spectrum of a matrix-free film containing CdTe QDs of ≈5 nm in size [56]. The spectrum of (non-absorbing) Si substrate is also shown for comparison. (b) Experimental RRS spectrum (excited with the 488 nm line of an Ar$^+$ laser) of CdTe NCs (4.5 nm in size) and its multi-Lorentzian fit showing the "anomalous" contribution of TO-type phonons [58] (reproduced with permission of the American Institute of Physics).

A recently published [58] resonant Raman scattering (RRS) spectrum of CdTe NCs is shown in Fig. 1.7b. It is similar to the RRS spectra of other II–VI NCs, pretty well understood and modeled (see, for example, [54]), except for one thing. There is a clear contribution of TO-type phonon modes [58], which is never seen, for example, for CdSe spherical QDs. Such an "anomalous" Raman mode has been observed only for InP dots (by several groups) and theoretically explained by the strong contribution of the optical deformation potential (ODP) mechanism of the exciton-phonon interaction in that material [54,59]. Among the II–VI-s, the (usually overlooked) ODP mechanism is relevant only for small CdTe QDs because of the high ODP constant (22.8 eV [60]) characteristic of bulk cadmium telluride. The results of recent calculations of RRS spectra of CdTe spherical dots taking into account the ODP mechanism, along with the dominating Fröhlich one, nicely agree with this experimental spectrum [59].

### 1.2.4 Up-Converted PL

Here we briefly discuss a phenomenon involving both quantum dot excitons and phonons. When an ensemble of QDs is excited at the lower edge of its absorption spectrum, a PL up-conversion effect can take place [61], which is the emission of photons with energies, $\hbar\Omega_E$, higher than the incident photon energy, $\hbar\Omega_I$. Along with InP NCs and CdSe/ZnS core/shell nanoparticles, it has been observed for colloidal CdTe QDs [33–35] (see Fig. 1.8). The principal experimental facts concerning the anti-Stokes shifted photoluminescence (ASPL) effect are the following [61]:

(i) The emission intensity increases linearly with the excitation power, which can be rather low.
(ii) The blue (i.e., anti-Stokes) shift between the PL peak energy, $\hbar\Omega_E - \hbar\Omega_I > 0$, does not significantly depend on the average QD size if $\hbar\Omega_I$ is chosen proportionally to the absorption peak energy, $E_{ex}$ (which does depend on the size). However, the shift always increases with temperature and can range from 20 to 150 meV.
(iii) If $\hbar\Omega_I$ increases approaching $E_{ex}$, the ASPL emission peak also moves continuously towards higher energies. Its intensity increases and finally the spectrum transforms into the normal PL band.
(iv) The ASPL intensity increases strongly with temperature.

The universal nature of this effect hints that the up-conversion mechanism should also be universal. It requires intermediate states (to which the electron is excited when the photon energy is not sufficient for an inter-band transition) and these states should be intrinsic to the dot. For example, assuming that the intermediate energy levels are related to some surface defects (as suggested in [34,35]), it would be hard to explain the similarity of the up-conversion effect in QDs of different size and even in different nanoparticle materials. On the other hand, as known, any discrete level of an electron or an exciton interacting with optical phonons gives rise to a ladder of energy states of a composite particle called a polaron. Therefore it was proposed [62] that the intermediate levels required for the up-conversion are just the exciton–polaron states lying below the ground exciton level (i.e., zero-phonon polaron line). After being excited through one of these states, the dot can emit a photon, most likely with the frequency corresponding to the zero-phonon line (ZPL) as shown in the inset of Fig. 1.8. The energy gain of (roughly) $\hbar\omega_{LO}$ comes from the thermostat that maintains the dot at a given temperature.

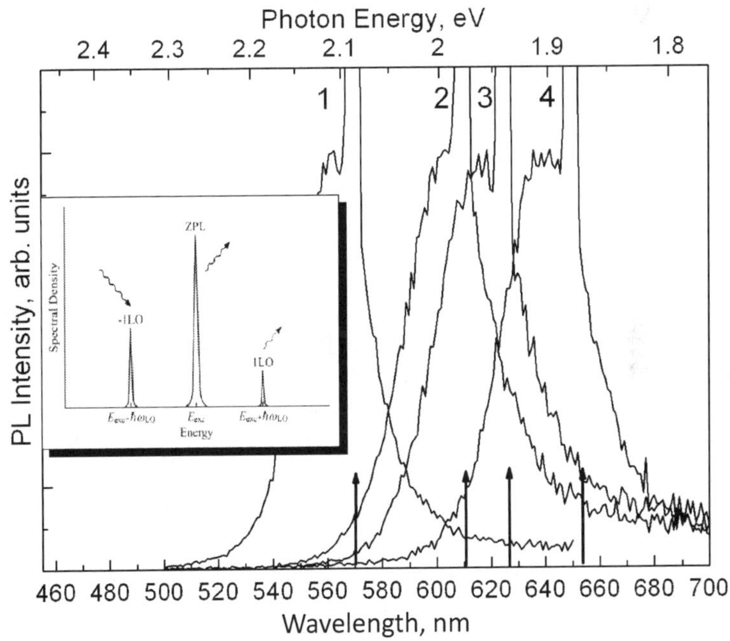


**Figure 1.8** Up-converted PL of CdTe QDs of different sizes in solution, 1.8 nm (1), 3.0 nm (2), 3.5 nm (3), and 4.4 nm (4) [33]. The inset shows the elementary processes responsible for the up-conversion in a single QD, absorption through a sub-gap polaron state and emission from the zero-phonon line or a higher state.

This explains why the anti-Stokes shift (i.e., the gap between the intermediate and emitting states) does not depend on the QD radius [61]. Spectral shifts larger than $\hbar\omega_{LO}$ can be explained by a cascade excitation as described in [39,63]. Indeed, the ASPL is excited through the largest (and rare) dots in the ensemble. In some cases, the emission occurs at a higher energy and such a photon has a larger chance to be re-absorbed because there are more dots of the appropriate size. Of course, many of the emitted photons just leave the sample contributing to the "normal" (unshifted or slightly Stokes-shifted) PL. The cascade re-absorption and re-emission process continues until the number of dots able to re-absorb becomes too small. Then, beyond the "normal" luminescence, a spectral maximum is formed at a higher frequency. Monte-Carlo modeling of these processes, performed in [63] confirms the experimental trends (i–iv). So far, the ASPL effect has been observed only for CdTe QDs dispersed in a colloidal solution (with a sufficiently high optical density). In principle, it should also occur in NC films where the cascade up-conversion could proceed through the Förster resonant energy transfer (FRET) mechanism. It would be interesting, in particular, for solid-state cooling [64].

## 1.3 Summary

We have given an overview of the optical properties of bulk and nanocrystalline cadmium telluride in connection with the underlying electron and phonon states and we have shown how these states in CdTe QDs are related to their bulk counterparts. From the theoretical viewpoint, it is an exemplary QD material. Because of the large exciton Bohr radius, the electrons and holes in these dots are in the strong confinement regime, with the exciton lifetime of the order of tens of nanoseconds [66]. The valence and conduction bands of CdTe are offset from those of CdSe and CdS, so that it is possible to prepare type-II core/shell NCs from combinations of these materials. Thanks to the cubic crystal structure and direct bandgap of the material, the QD exciton and phonon states can be described by simple models and it makes their control easier. In contrast with other II–VI materials, the exciton-phonon interaction in QDs occurs *via* both the Fröhlich and ODP mechanisms and the latter reveals itself by the presence of a TO-type mode in the Raman spectra. The efficient exciton interaction with a broader variety of phonon modes (of both LO and TO-types) results in a denser exciton–polaron spectrum, which should contribute to the polaron relaxation from an excited state to the lowest energy one, thus leading to a faster optical response of CdTe dots, compared to other QD materials. It should also enhance the resonant energy transfer processes in QD ensembles and hybrid-QD systems by enlarging the energy matching conditions.

## References

1. Zanio, K. (1978) *Semiconductors and Semimetals* **13**, eds. Willardson, R. K., and Beer, A. C., "Cadmium Telluride" (Academic Press, New York).
2. Collaboration: Authors and editors of the volumes III/17B-22 A-41B (1999) eds. Madelung, O., Rössler, U., Schulz, M., *Springer Materials—The Landolt–Börnstein Database* (http://www.springermaterials.com).
3. Segall, B., Lorenz, M. R., and Halsted, R. E. (1963). Electrical properties of n-type CdTe, *Phys. Rev.*, **129**, pp. 2471–2481.
4. Jimenez Sandoval, S., Meléndez Lira, M., and Hernández Calderón, I. (1992). Crystal structure and energy gap of CdTe thin films grown by radio frequency sputtering, *J. Appl. Phys.*, **72**, pp. 4197–4202.
5. Nelmes, R. J., McMahon, M. I., Wright, N. G., and Allan, D. R. (1995). Phase transitions in CdTe to 28 GPa, *Phys. Rev. B*, **51**, pp. 15723–15731.
6. Chelikowsky, J. R., and Cohen, M. L. (1976). Nonlocal pseudopotential calculations for the electronic structure of eleven diamond and zinc-blende semiconductors, *Phys. Rev. B*, **14**, pp. 556–582.
7. Cade, N. A., and Lee, P. M. (1985). Self consistent energy band structures for HgTe and CdTe, *Solid State Commun.*, **56**, pp. 637–641.
8. Zakharov, O., Rubio, A., Blase, X., Cohen, M. L., and Louie, S. G. (1994). Quasiparticle band structures of six II–VI compounds: ZnS, ZnSe, ZnTe, CdS, CdSe, and CdTe, *Phys. Rev. B*, **50**, pp. 10780–10787.
9. Dunstan, D. J., Gil, B., Priester, C., and Homewood, K. P. (1989). Hydrostatic pressure dependence of CdTe, *Semicond. Sci. Tech.*, **4**, pp. 241–249.
10. Nicklas, J. W., and Wilkins, J. W. (2011). Accurate electronic properties for (Hg, Cd)Te systems using hybrid density functional theory, *Phys. Rev. B*, **84**, pp. 121308-1–4.

11. cWard, D. K., Zhou, X. W., Wong, B. M., Doty, F. P., and Zimmerman, J. A. (2011). Accuracy of existing atomic potentials for the CdTe semiconductor compound, *J. Chem. Phys.*, **134**, pp. 244703-1–13.
12. Marple, D. T. F., and Ehrenreich, H. (1962). Dielectric constant behavior near band edges in CdTe and Ge, *Phys. Rev. Lett.*, **8**, pp. 87–89.
13. Cardona, M., and Greenaway, D. L. (1963). Fundamental reflectivity and band structure of ZnTe, CdTe, and HgTe, *Phys. Rev.*, **131**, pp. 98–103.
14. Niles, D. W., and Hochst, H. (1991). Critical test of CdTe (100) angle-resolved photoemission spectra with band-structure calculations, *Phys. Rev. B*, **43**, pp. 1492–1499.
15. Nikitin, P. I., and Savchuk, A. I. (1990). The Faraday effect in semi-magnetic semiconductors, *Sov. Phys. Uspekhu*, **33**, pp. 974–1005.
16. Lee, J., Giles, N. C., Rajavel, D., and Summers, C. J. (1994). Room-temperature band-edge photoluminescence from cadmium telluride, *Phys. Rev. B*, **49**, pp. 1668–1676.
17. Horodyský, P., and Hlídek, P. (2006). Free-exciton absorption in bulk CdTe: temperature dependence, *Phys. Stat. Sol. (b)*, **243**, pp. 494–501.
18. Pässler, R. (2003). Semi-empirical descriptions of temperature dependences of band gaps in semiconductors, *Phys. Stat. Sol. (b)*, **236**, pp. 710–728.
19. Varshni, Y. P. (1967). Temperature dependence of the energy gap in semiconductors, *Physica*, **34**, pp. 149–154.
20. Eech, V., *et al.* (1990). State filling, Coulomb, and trapping effects in the optical nonlinearity of CdTe quantum dots in glass, *Phys. Rev. B*, **42**, pp. 7450–7453.
21. Bryant, F. J., P., Totterdell, D. H. J., and Hagston, W. E. (1971). Identification of edge and exciton emission centres in CdTe, *J. Phys. C: Solid State Phys.*, **4**, pp. 641–649.
22. Yu, P. Y., and Cardona, M. (1996). *Fundamentals of Semiconductors* (Springer-Verlag, Berlin).
23. Brazis, R., Nausewicz, D., and Raguoti, R. (2007). Phonon- and electron-related far-infrared absorption in CdTe and ZnTe crystals, *Phys. Stat. Sol. (b)*, **244**, pp. 1662–1668.
24. Dal Corso, A., Baroni, S., and Resta, R. (1993). *Ab initio* calculation of phonon dispersions in II–VI semiconductors, *Phys. Rev. B*, **47**, pp. 3588–3592.

25. Rajput, B. D., and Browne, D. A. (1996). Lattice dynamics of II–VI materials using the adiabatic bond-charge model, *Phys. Rev. B*, **53**, pp. 9052–9058.

26. Rowe J. M., Nicklow, R. M., Price, D. L., and Zanio, K. (1974). Lattice dynamics of cadmium telluride, *Phys. Rev. B*, **10**, pp. 671–675.

27. Vasilevskiy, M. I., Akinkina, E. I., de Paula, A. M., and Anda, E. V. (1998). Effect of size dispersion on the optical absorption of an ensemble of semiconductor quantum dots, *Semiconductors*, **32**, pp. 1229–1233.

28. Murray, C. B., Norris, D. J., and Bawendi, M. G. (1993). Synthesis and characterization of nearly monodisperse CdE (E = S, Se, Te) semiconductor nanocrystallites, *J. Am. Chem. Soc.*, **115**, pp. 8706–8715.

29. Rajh, T., Micic, O. I., and Nozik, A. J. (1999). Synthesis and characterization of surface-modified colloidal CdTe quantum dots, *J. Phys. Chem.*, **97**, pp. 11999–12003.

30. Gaponik, N., and Rogach, A. L. (2008). *Semiconductor Nanocrystal Quantum Dots*, ed. Rogach A. L., Chapter 3, "Aqueous synthesis of semiconductor nanocrystals" (Springer-Verlag, Wien), pp. 73–99.

31. Rockenberger, J., Tröger, L., Rogach, A. L., Tischer, M., Grundmann, M., Eychmüller, A., and Weller, H. (1998). The contribution of particle core and surface to strain, disorder and vibrations in thiolcapped CdTe nanocrystals, *J. Chem. Phys.*, **108**, pp. 7807–7815.

32. Hollingsworth, J. A., and Klimov, V. I. (2010). *Nanocrystal Quantum Dots*, ed. Klimov, V. I., Chapter 1, "Soft chemical synthesis and manipulation of semiconductor nanocrystals" (CRC Press, Boca Raton, FL, USA), pp. 1–62.

33. Filonovich, S. A. (2003). *Optoelectronic Properties of Semiconductor Quantum Dot Ensembles: The Role of Electron–Phonon Interaction and Collective Effects* (PhD Thesis, University of Minho, Braga, Portugal).

34. Rakovich, Y. P., Gladyshchuk, A. A., Rusakov, K. I., Filonovich, S. A., Gomes, M. J. M., Donegan, J. F., Talapin, D. V., Rogach, A. L., and Eychmüller, A. (2002). Anti-Stokes luminescence of cadmium telluride nano-crystals, *Appl. Spectrosc.*, **69**, pp. 444–449.

35. Wang, X., Yu, W. W., Zhang, J., Aldana, J., Peng, X., and Xiao, M. (2003). Photoluminescence up-conversion in colloidal CdTe quantum dots, *Phys. Rev. B*, **68**, pp. 125318-1–6.

36. Efros, Al. L., Rosen, M., Kuno, M., Nirmal, M., Norris, D. J., and Bawendi, M. (1996). Band-edge exciton in quantum dots of semiconductors with a degenerate valence band: dark and bright exciton states, *Phys. Rev. B*, **54**, pp. 4843–4856.

37. Tan, G. L., Yang, Q., Hömmerich, U., Seo, J. T., and Temple, D. (2004). Linear and non-linear optical properties of capped CdTe nanocrystals prepared by mechanical alloying, *Opt. Mater.*, **27**, pp. 579–584.

38. Padilha, L. A., Neves, A. A. R., Rodriguez, E., Cesar, C. L., Barbosa, L. C., and Brito Cruz, C. H. (2005). Ultrafast optical switching with CdTe nanocrystals in a glass matrix, *Appl. Phys. Lett.*, **86**, pp. 161111-1–3.

39. Vasilevskiy, M. I. (2008). *Semiconductor Nanocrystal Quantum Dots*, ed. Rogach A. L., Chapter 8, "Exciton–phonon interaction in semiconductor nanocrystals" (Springer-Verlag, Wien), pp. 257–275.

40. Brus, L. E. (1984). Electron-electron and electron–hole interactions in small semiconductor crystallites: The size dependence of the lowest excited electronic state, *J. Chem. Phys.* **80**, pp. 4403–4408.

41. Sercel, P. C., and Vahala, K. J. (1990). Analytical formalism for determining quantum-wire and quantum-dot band structure in the multiband envelope-function approximation, *Phys. Rev. B*, **42**, pp. 3690–3710.

42. Efros, Al. L. (2010). *Nanocrystal Quantum Dots*, ed. Klimov, V. I., Chapter 3 "Fine structure and polarization properties of band-edge excitons in semiconductor nanocrystals" (CRC Press, Boca Raton, FL, USA), pp. 97–132.

43. Delerue C., and Lannoo M. (2004). *Nanostructures. Theory and Modelling* (Springer-Verlag, Berlin).

44. Jasieniak, J., Califano, M., and Watkins, S. E. (2011). Size-dependent valence and conduction band-edge energies of semiconductor nanocrystals, *ACS Nano*, **5**, pp. 5888–5902.

45. Pellegrini, G., Mattei, G., and Mazzoldi, P. (2005). Finite depth square well model: Applicability and limitations, *J. Appl. Phys.*, **97**, pp. 073706-1–5.

46. Baskoutas, S., and Terzis, A. F. (2006). Size-dependent gap of colloidal quantum dots, *J. Appl. Phys.*, **99**, pp. 013706-1–4.

47. Prado, S. J., Trallero-Giner, C., Alcalde, A. M., López-Richard, V., and Marques, G. E. (2003). Optical transitions in a single CdTe spherical quantum dot, *Phys. Rev. B*, **68**, pp. 235327-1–9.

48. Núñez Fernández, Y., Vasilevskiy, M. I., Larramendi, E. M., and Trallero-Giner, C. (2012). *Quantum Dots/Book I*, ed. Ameenah, A.-A., Chapter X, "Exciton states in free-standing and embedded semiconductor nanocrystals" (Intech Open Publisher, Rijeka).

49. Yu, W. W., Qu, L., Guo, W., and Peng, X. (2003). Experimental determination of the extinction coefficient of CdTe, CdSe, and CdS nanocrystals, *Chem. Mater.*, **15**, pp. 2854–2860.

50. Rogach, A. L., Franzl, T., Klar, T. A., Feldmann, J., Gaponik, N., Lesnyak, V., Shavel, A., Eychmüller, A., Rakovich, Yu. P., and Donegan, J. F. (2007). Aqueous synthesis of thiol-capped CdTe nanocrystals: State-of-the-art, *J. Phys. Chem. C*, **111**, pp. 14628–14637.

51. Haram, S. K., Kshirsagar, A., Gujarathi, Y. D., Ingole, P. P., Nene, O. A., Markad, G. B., and Nanavati, S. P. (2011). Quantum confinement in CdTe quantum dots: Investigation through cyclic voltammetry supported by density functional theory (DFT), *J. Phys. Chem. C*, **73**, pp. 6243–6249.

52. Vasilevskiy, M. I., and Anda, E. V. (1996). Effective dielectric response of semiconductor composites, *Phys. Rev. B*, **54**, pp. 5844–5851.

53. Stroscio, M. A., and Dutta, M. (2001). *Phonons in Nanostructures* (Cambridge University Press, Cambridge).

54. Rolo, A. G., and Vasilevskiy, M. I. (2007). Raman spectroscopy of optical phonons confined in semiconductor quantum dots and nanocrystals, *J. Raman Spectrosc.*, **38**, pp. 618–633.

55. Bragas, A. V., Aku-Leh, C., and Merlin, R. (2006). Raman and ultrafast optical spectroscopy of acoustic phonons in $CdTe_{0.68}Se_{0.32}$ quantum dots, *Phys. Rev. B*, **73**, pp. 125305-1–9.

56. Rolo, A. G., Vasilevskiy, M. I., Gaponik, N. P., Rogach, A. L., and Gomes, M. J. M. (2002). Confined optical vibrations in CdTe quantum dots and clusters, *Phys. Stat. Sol. (b)*, **229**, pp. 433–437.

57. Vasilevskiy, M. I., (2002). Dipolar vibrational modes in spherical semiconductor quantum dots, *Phys. Rev. B*, **66**, pp. 195326-1–9.

58. Dzhagan, V., Valakh, M. Ya., Kolny-Olesiak, J., Lokteva, I., and Zahn, D. R. T. (2009). Resonant Raman study of phonons in high-quality colloidal CdTe nanoparticles, *Appl. Phys. Lett.*, **94**, pp. 243101-1–3.

59. Vasilevskiy, M. I., and Trallero-Giner, C. (2010). Resonant Raman scattering in spherical QDs: II–VI versus III–V semiconductor nanocrystals, *Phys. Stat. Sol. (b)*, **247**, pp. 1488–1491.

60. Blacha A., Presting H., and Cardona M. (1984). Deformation potentials of $k = 0$ states of tetrahedral semiconductors, *Phys. Stat. Sol. (b)*, **126**, pp. 11–36.

61. Rakovich, Y. P., and Donegan, J. F. (2008). *Semiconductor Nanocrystal Quantum Dots*, ed. Rogach A. L., Chapter 9, "Anti-Stokes

photoluminescence in semiconductor nanocrystal quantum dots" (Springer-Verlag, Wien), pp. 257–275.

62. Vasilevskiy, M. I., Anda, E. V., and Makler, S. S. (2004). Electron–phonon interaction effects in semiconductor quantum dots: a non-perturbative approach, *Phys. Rev. B*, **70**, pp. 035318-1–14.
63. Santos, J. R., Vasilevskiy, M. I., and Filonovich, S. A. (2008). Cascade upconversion of photoluminescence in quantum dot ensembles, *Phys. Rev. B*, **78**, pp. 245422-1–6.
64. Rakovich, Yu. P., Donegan, J. F., Vasilevskiy, M. I., and Rogach, A. L. (2009). Anti-Stokes cooling in semiconductor nanocrystal quantum dots: A feasibility study, *Phys. Stat. Sol. (a)*, **206**, pp. 2497–2510.
65. Rogach, A. L. (2000). Nanocrystalline CdTe and CdTe(S) particles: wet chemical preparation, size-dependent optical properties and perspectives of optoelectronic applications, *Mater. Sci. Eng. B*, **69–70**, pp. 435–440.
66. Yan, Y., Chen, G., and Van Patten, P. G. (2011). Ultrafast exciton dynamics in CdTe nanocrystals and core/shell CdTe/CdS nanocrystals, *J. Phys. Chem. C*, **115**, pp. 22717–22728.

## Chapter 2

# Aqueous Synthesis of Colloidal CdTe Nanocrystals

**Vladimir Lesnyak,**[a,b] **Nikolai Gaponik,**[a] **and Alexander Eychmüller**[a]

[a]*Physical Chemistry, Technical University of Dresden, 01062 Dresden, Germany*
[b]*Istituto Italiano di Tecnologia, Via Morego 30, 16163 Genova, Italy*

vladimir.lesnyak@chemie.tu-dresden.de

## 2.1 Introduction

Colloidal aqueous synthesis of semiconductor nanocrystals (NCs), also referred to as quantum dots (QDs), is an advantageous alternative to the widely used organometallic route [1]. Compared with that route, the aqueous approach (1) does not require high temperatures and glove boxes, (2) employs the most widespread and biocompatible solvent—water, (3) is easily up-scalable up to industrial requirements, and (4) provides various functionalization of NCs *via* applying an appropriate capping ligand, which in turn may be further functionalized by electrostatic or covalent linking. This method allows successful synthesis of a series of various semiconductor nanoparticulate materials directly in water

*Cadmium Telluride Quantum Dots: Advances and Applications*
Edited by John F. Donegan and Yury P. Rakovich

ISBN 978-981-4316-05-7 (Hardcover), 978-981-4316-09-5 (eBook)
www.panstanford.com

employing mild preparative conditions. In general, the aqueous synthetic protocol consists of the reaction between metal and chalcogen containing precursors in the presence of an appropriate stabilizer with subsequent nucleation and growth of NCs, typically by heat treatment. In the case of aqueous synthesis, thiols are definitely the most used ligands for the stabilization of various semiconductor materials [2]. The formation of nanoparticles is a dynamic process, which is usually explained by an Ostwald ripening (OR) mechanism mainly consisting of the growth of larger particles at the expense of smaller ones present in an ensemble. The history of this method extends over 30 years beginning from an early work of Henglein's group [3]. For a comprehensive historical overview of progress made in the field during this time, the reader may find these recent publications useful [4–7].

Although the first material synthesized *via* an aqueous approach and thoroughly investigated was CdS, the most successful example of the cadmium chalcogenides is CdTe owing to its unique optical properties, *e.g.*, strong photoluminescence (PL) in the visible region. CdTe NCs had been obtained directly in water first by Resch *et al.* through the reaction of cadmium perchlorate with sodium telluride in the presence of sodium hexametaphosphate as stabilizer [8]. Rajh *et al.* employed a mixed stabilizer system containing hexametaphosphate and 3-mercapto-l,2-propanediol (thioglycerol (TG)) for successful synthesis of CdTe NCs [9]. Later, optimization by Rogach *et al.* led to a synthesis of stable NCs with sizes from 1.3 to 2.4 nm in the sole presence of thiols, *viz.*, 2-mercaptoethanol and 1-thioglycerol [10]. State-of-the-art preparation techniques enable facile aqueous synthesis of CdTe QDs capped by different thiols, possessing diameters of up to 6 nm and PL covering most part of the visible region extending to the near infrared (NIR) with quantum yields (QYs) of up to 70–80%, which can compete with the best organometallicaly prepared materials [11]. For instance, as-prepared NCs by microwave irradiation demonstrate PL QY as high as 82% and a narrow size distribution [12]. Here we would like to note that very high QYs reported should be considered very carefully, since the QY value is very much dependent on the method of determination, standards used, instrument conditions, and so on. A reliable and quite detailed procedure for QY measurement has been reported recently by Grabolle *et al.* [13]. Luminescence of CdTe NCs determines their potential applicability in the fields of light emitting diodes (LEDs) fabrication [14,15], color conversion [16], energy scavenging [17,18], optical sensing [19] and bioimaging [20].

Based on the synthesis of CdTe nanoparticles, approaches for obtaining some alloyed materials such as CdSeTe [21], CdHgTe [22–24], as well as core/shell structures such as CdTe/CdS [25–27], CdTe/ZnTe [28], and even core/shell/shell CdTe/CdS/ZnS [29] have been developed. The obtaining of different core/shell and core/shell/shell structures directly in water reported during the last decade is a great step forward toward a superior quality of NCs accessible from an organometallic route. Interesting hybrid approaches for the synthesis of core/shell particles combining aqueous and organometallic techniques were developed recently. Thereby CdTe and CdHgTe cores transferred from water were coated by ZnS shell *via* the hot injection technique [30]. As a recent advance, a facile synthesis of 1D structures such as CdTe [31,32] and CdHgTe [33] nanorods, allowing control of their sizes and aspect ratios, should be mentioned. Nevertheless, we have to admit that the aqueous method usually does not permit direct efficient shape control of nanoparticles during their growth due to temperature limitations: temperature of around 100°C is not sufficient to overcome the energy barrier of transforming zinc blende to wurtzite structure. Thus, NCs synthesized exhibit as a rule an isotropic zinc blende cubic crystal structure and mainly spherical or quasi-spherical shape, since this shape is thermodynamically the most stable [34]. Therefore, shaped CdTe nanostructures, such as 1D wires [35,36], twisted ribbons [37], 2D sheets [38], 3D gel-like networks [39,40] are achievable as a rule *via* postpreparative self-organization of preformed QDs. Assembly approaches of aqueous colloidal CdTe QDs are discussed in Chapter 3.

Mild synthetic conditions used in an aqueous synthesis often disable perfect structural localization of atoms in the crystal lattice of nanoparticles, normally achievable through the hot injection method. In order to improve crystal structure, size distribution and correspondingly optical properties of CdTe QDs, microwave irradiation has been successfully applied [12,25,29,41,42].

In the following paragraphs, we will review in detail the following aspects of the aqueous synthesis of CdTe NCs: its synthetic protocol, growth mechanism, postpreparative treatment of the particles, and obtaining some alloyed and core/shell structures based on the CdTe matrix.

## 2.2 Synthetic Protocol

The typical synthetic protocol for obtaining aqueous thiol-capped CdTe NCs is a one-pot process consisting of three main steps:

(1) Formation of $Cd^{2+}$/thiol complexes by dissolving a Cd salt (usually $Cd(ClO_4)_2$ or $CdCl_2$) in water followed by addition of an appropriate thiol with subsequent pH adjustment. Note, cysteamine stabilization does not require pH adjustment.

(2) Injection of a $Te^{2-}$ source ($H_2Te$ gas, NaHTe) in a thoroughly de-aerated (through purging by an inert gas (Ar, $N_2$)) $Cd^{2+}$/thiol solution results in the formation of a CdTe precursor (often referred as a monomer). Hydrogen telluride is produced either by decomposition of $Al_2Te_3$ (see Fig. 2.1) or electrochemical generation, a detailed recipe is reported in ref. [11].

(3) Nucleation and growth of NCs by heating or microwave irradiation, usually upon reflux under open air conditions.

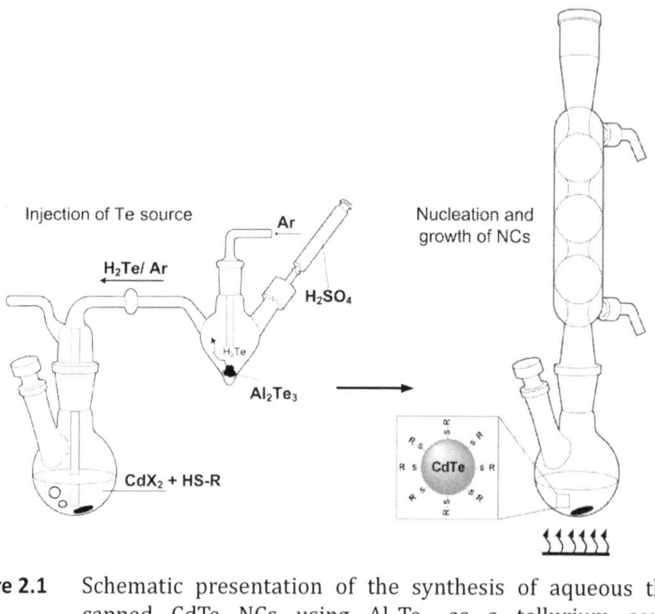


**Figure 2.1** Schematic presentation of the synthesis of aqueous thiol-capped CdTe NCs using $Al_2Te_3$ as a tellurium source. Courtesy of A. Dubavik (LMU Munich, Germany).

A variety of thiol capping ligands employed in the aqueous synthesis of CdTe and other thiol-stabilized NCs is summarized in Fig. 2.2. Those thiols possessing optical isomerism, like L- and D-cysteine, L- and D-cysteinemethylester hydrochloride, bidentate R- and S-α-lipoic acids impart chiral structure to the CdTe surface [43,44]. The resulting NCs keep their optical activity even after ligand exchange for an achiral stabilizer (so-called chiral memory effect).

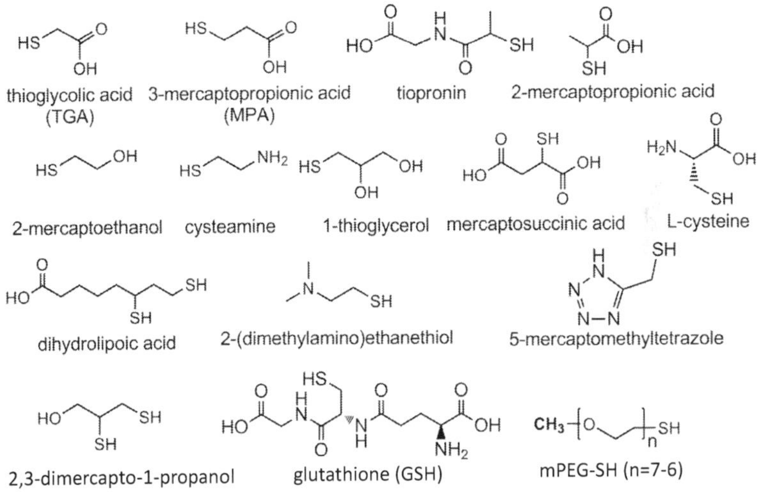


**Figure 2.2** Chemical structures of typical thiol capping ligands employed in the aqueous synthesis of CdTe NCs. Adapted with permission from ref. [2]. Copyright 2010, The Royal Society of Chemistry.

Figure 2.3 shows typical PL and absorbance spectra of thioglycolic acid (TGA) and 3-mercaptopropionic acid (MPA) capped CdTe QDs along with the evolution of PL lifetime of CdTe/MPA NCs during their growth and the transmission electron microscopy (TEM) image of the CdTe/MPA sample emitting with a maximum at 780 nm. We note that MPA stabilization ensures very fast growth [45] of the largest achievable NCs still possessing emission peaking at 840 nm [22], which is quite close to the bulk bandgap of CdTe (1.56 eV at 300 K) [9]. Nevertheless, the Bohr radius of 7.3 nm [9] still provides efficient quantization of energy levels within the particle.

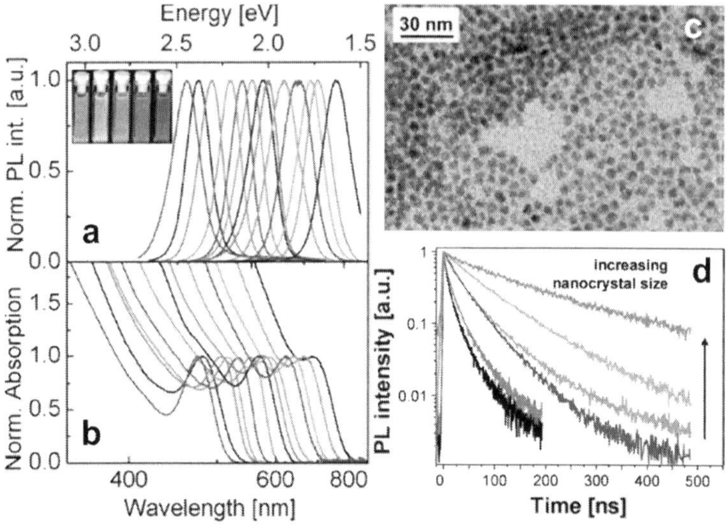


**Figure 2.3** PL (a) and absorption (b) spectra of TGA- and MPA-capped CdTe NCs demonstrate their tuneability over a broad spectral range in the visible and near-infrared ($\lambda_{ex}$ = 450 nm). TEM image of MPA-capped CdTe NCs, 5.5 nm average size, with a PL maximum at 780 nm (c). PL decays of MPA-capped CdTe NCs of increasing sizes (d). Adapted with permission from ref. [11]. Copyright 2007, American Chemical Society.

Aqueous CdTe NCs display different stability in varied water based media, which strongly depends on their size and concentration. Thus, Boldt *et al.*, having tested TGA- and MPA-stabilized CdTe particles in various buffers (in the pH range 5–11) commonly used for the indirect immunofluorescence labeling of cells, showed that a lower concentration of QDs results in the detachment of ligands from the particle surface and therefore a decreased stability [46]. The control parameter being monitored during a period of one week was the PL of several CdTe NCs fractions derived *via* size-selective precipitation of the colloids (described in Section 2.4). As expected, large semiconductor QDs have shown a greater resistance to photobleaching than smaller ones. However, smaller QDs displayed higher stability under acidic conditions and low buffer concentrations; this was most probably due to their clusterlike, more defined structure. The authors have observed pseudozero-order kinetics of the emission diminution, ascribing it to the oxidation of the particles surface by the dissolved oxygen.

Although as it was already mentioned above, the aqueous method does not permit direct synthesis of the CdTe particles with a nonspherical shape, a few examples of the formation of 1D nanostructures during the NCs growth have been reported [31,32,47]. The key factors ensuring the CdTe growth into one direction and subsequent stability of the obtained structures are the low concentration of the precursors, employment of the mixed ligand system and moderate temperature in order to suppress OR of the NCs. Thereby, Zhang *et al.* successfully prepared CdTe nanorods using TGA and TG as a mixed stabilizer and incubating the precursor solution with concentration of $Cd^{2+}$ = 2.4 mM and a ratio of $Cd^{2+}/TGA/TG/HTe^{-}$ = 1/1.2/1.2/0.5 at 80°C for several hours (see Fig. 2.4) [32,47]. In the presence of solely TGA,

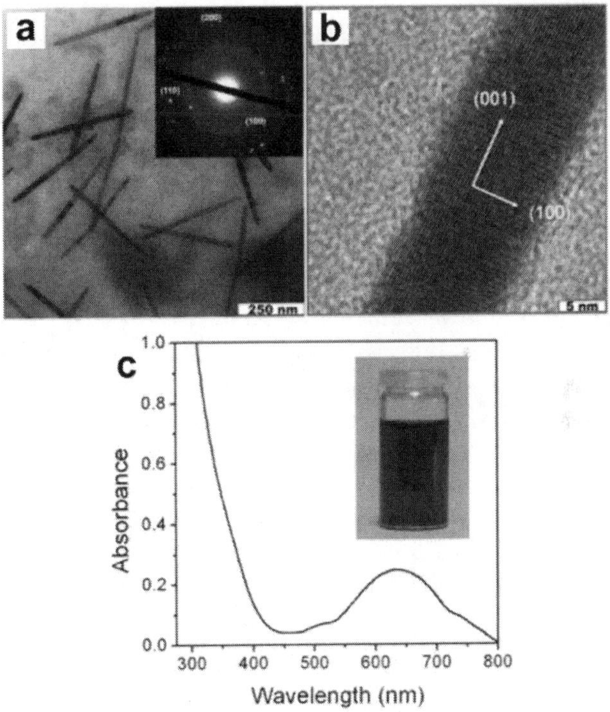


**Figure 2.4** TEM (a) and high-resolution TEM (HRTEM) (b) images of 1D CdTe nanocrystals obtained by 8 h storage of 2.4 mM precursor solutions in the presence of TGA and TG at 80°C. Inset in (a) is their SAED pattern. Corresponding absorbance spectrum and the photograph (c). Reproduced with permission from ref. [47]. Copyright 2007, American Chemical Society.

metastable 1D wurtzite structures under present conditions have transformed into thermodynamically stable spherical zinc blende NCs. Furthermore, applying the mixture of TGA and 2-mercaptoethylamine allows for a control of the aspect ratio of CdTe nanorods during their growth [47]. The elongation may cause a long-wavelength shift of the optical absorption providing the resulting dark color of the nanorods suspension (see Fig. 2.4c). These nanorods exhibit a weak luminescence owing to the weaker quantum confinement of 1D nanostructures, relative to that of 0D QDs.

## 2.3 CdTe Nanocrystal Growth Mechanism

Similar to the organometallic route, the formation of aqueous NCs consists of the fast nucleation stage followed by slow OR. Coalescence of the precursors—monomers and even some magic clusters—is the main process accounting for nucleation, whereas OR comes to the foreground after monomer depletion. It is also considered that an oriented attachment (OA) mechanism coexists simultaneously with OR, especially at the early stage of the particles growth, when two crystallographically oriented nanoparticles combine together to form a larger one [48].

Owing to the lower temperature, different precursor reactivity, and reaction media, the NC growth rate is much slower than that in the organometallic approach, and therefore the formation of aqueous NCs has its own peculiarities. The kinetics of the NC growth is determined mainly by the concentration of precursors, the nature of a stabilizer, $Cd^{2+}/Te^{2-}$/thiol ratio, pH of the solution, and its temperature. Yang *et al.* explained the growth of CdTe NCs by electrostatic effects occurring on the nanoparticle surface [49]. According to this suggestion, the nature of the diffuse layer influences the ability of the monomer (*i.e.*, CdTe precursor) diffusion and hence the growth rate of NCs (see Fig. 2.5). Thus, the reaction conditions reducing the thickness of the electric double-layer benefit monomer diffusion and a rapid growth of NCs. At the same time, the diffuse layer also regulates a charge-selective transfer of Cd monomers. The neutral monomers, such as the complex Cd(MPA), migrate through the diffuse layer easier than the charged ones $Cd(MPA)_2^{2-}$ or $Cd(MPA)_3^{4-}$, thus facilitating the growth of NCs. Electrostatic effects play an important role not only in the monomer diffusion but also in the interparticle interaction, especially at the initial stage of the reaction, where CdTe NCs grow through agglomeration of small clusters [50]. At this step decrease of the electrostatic repulsion facilitates aggregation and consequently growth of the particles.

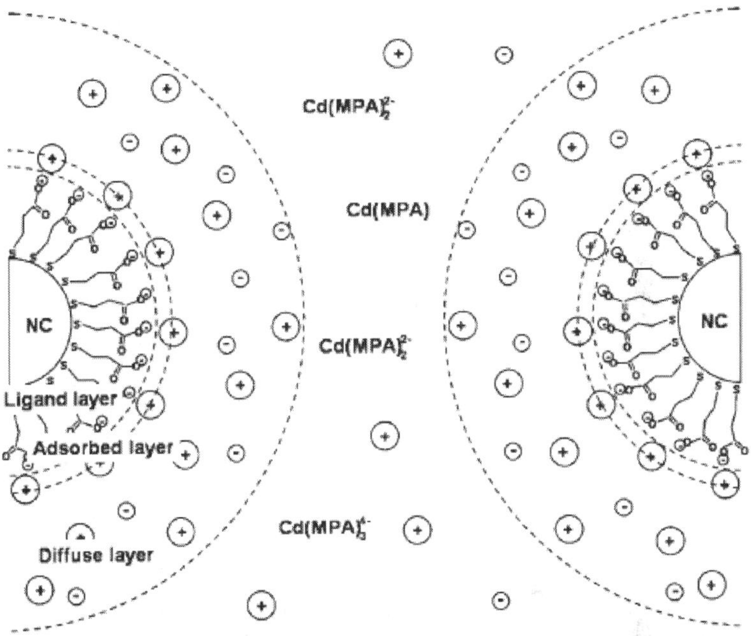


**Figure 2.5** Schematic illustration of the electric double-layer of aqueous MPA-capped CdTe NCs. The structure consists of CdTe core, ligand layer, adsorbed layer and diffuse layer. The adsorbed layer corresponds to the contact adsorbed ions, which is only an instantaneous state for a growing NC. The adsorbed cation is predominantly $Na^+$. Reproduced with permission from ref. [49]. Copyright 2008, WILEY VCH Verlag GmbH & Co. KGaA.

Experimental results and calculations employing stability constants of cadmium complexes with TGA performed by Shavel *et al.* also reveal the domination of the uncharged Cd(TGA) complex in the NC growth, gaining further insight into the factors governing the NC growth [51]. A numerical calculation shows a clear correlation between the concentration of $Cd^{2+}$/thiol in the initial solution and the PL QY of the resulting CdTe NCs. An improvement of the emission efficiency of the CdTe QDs up to ca. 50–60% was achieved through decreasing the Cd/TGA ratio to 1:1.3, which is advantageous for increasing the concentration of the $Cd^{2+}$/thiol complex. Moreover, maintaining the solution pH at 12 allows the reduction of the full width at half-maximum (FWHM) and the Stokes shift of the PL. The decomposition of the Cd/thiol complexes rather than the diffusion of monomers is suggested as a reaction rate-limiting process [52]. In this respect, the stability of the Cd (MPA) complex is lower than that of Cd(TGA), which leads to a faster release of the monomers and correspondingly to a faster particle growth. The key role of a ligand shell in the growth process and PL QYs of aqueous CdTe NCs has also been proven considering its influence on the activation energy *via* formation of the transition-state complex [53].

As was proposed by Yin *et al.*, coalescence of small particles plays a major role at the initial growth stage of TGA-capped CdTe NCs, while its later period is mainly controlled by OR [48]. As seen in the TEM images presented in Fig. 2.6a–d, after 20 min of the reaction most of the irregularly shaped NCs exhibit the characteristics of growth by the OA mechanism, such as the retained lattice defects (twins and stacking faults) and clear coalescence profile of small building blocks, whereas at 200 min the particles evolved into a relatively perfect lattice and a more regular shape with smooth edges, achieved *via* OR. Fitting the growth kinetics further confirms this assumption: the evolution of the particle size is better fitted employing an OA mechanism at the early stage (first 40 min), whereas OR theory provides a better convergence at the later stage (after 50 min) (see Fig. 2.6e). Combining these two mechanisms, the authors derived Eq. 2.1 that well fits the experimental results:

$$d = \frac{(d_0^n + K_{\mathrm{OR}}t)^{1/n}(2^{1/3}N_0k_{\mathrm{OA}}t + 1)}{N_0k_{\mathrm{OA}}t + 1}, \qquad (2.1)$$

where $d$ is the mean particle size at $t$, $d_0$ is primary particle size, $K_{\mathrm{OR}}$ is a temperature-dependent OR growth rate constant, $n$ is an exponent relevant to the coarsening mechanism (physically meaningful in the range of 1–5), $k_{\mathrm{OA}}$ is the OA rate constant, $N_0$ is the primary particle concentration.

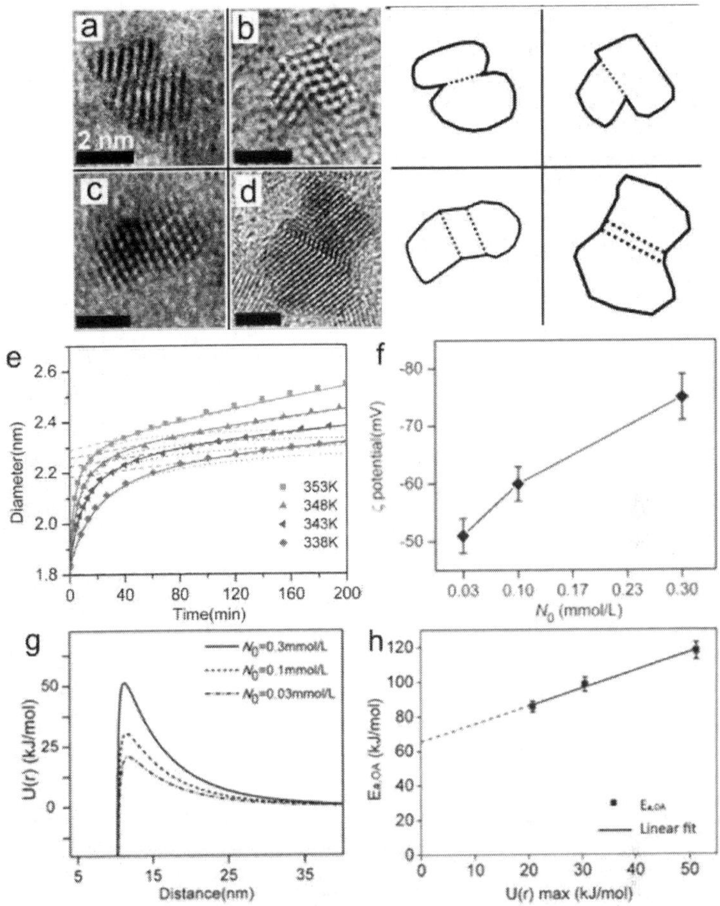


**Figure 2.6** (a–d) HRTEM images of defective CdTe NCs coarsened *via* OA and their corresponding simulations (right). (e) Evolution of the CdTe NCs sizes as a function of their growth time at different growth temperatures. The solid lines are obtained by fitting experimental results (dots) with OA and OR mixed growth model. The dotted and dashed lines represent the fitting by OA kinetic model and OR theory, respectively. (f) Zeta potentials of TGA-capped CdTe NCs at different primary particle concentrations. (g) Predicted electrostatic particle–particle interaction energy $U(r)$ between CdTe NCs in solutions with different particle concentrations $N_0$ = 0.3, 0.1 and 0.033 mM at $T$ = 338 K. (h) The OA activation energy ($E_{a,OA}$) versus the electrostatic barrier energy ($U(r)$ max). Adapted with permission from ref. [48]. Copyright 2011, American Chemical Society.

It was found that at the initial growth stage increasing particle concentration accelerates the growth rate providing more chances for collisions between NCs, while in the latter stage the particle growth becomes slower in the system with higher particle concentration. At the same time, as is seen from Figs. 2.6f–h, increase of the particles concentration leads to the rise of the surface charge density inhibiting OA by NCs repulsion and consequently results in a smaller OA rate constant and correspondingly higher activation energy. With the increase of particle sizes, the OA rate constant largely decreases due to the diffusion effect. In addition, in the course of the growth, the electrostatic particle–particle repulsion increases, and thus the OA rate constant decreases abruptly.

Strong interaction between the adsorbed layer inside the diffuse layer and Cd monomers provides the opportunity to accelerate NC growth and to obtain NCs with high PLQYs. This assumption elucidates the difference between the growth rates of CdTe capped by MPA and TG. Deprotonated carboxyl groups of MPA strongly interact with Cd assisting monomer adsorption. On the contrary, although similar amount of monomers transport through the diffuse layer formed by hydroxyl groups of TG, only a few monomers are adsorbed on the NC surface due to the weak interaction of Cd and –OH groups, retarding the growth. In general, a relatively quick particle growth leads to an insufficient quality reflected in low crystallinity and a large number of defects and surface states. Oppositely, a comparatively slow growth rate leads to a high content of sulfur (as a product of the thiol decomposition) in the particles and a higher probability of the NC oxidation [51]. It should be noted that it is impossible to completely prevent the incorporation of sulfur into the large nanocrystals either by the choice of the solvent or by a variation of the stabilizers [54]. Nevertheless, this process leads to the formation of a CdS layer on the CdTe NC surface (or CdTe(S) NCs with sulfur enriched shell) resulting in an improvement of the material's quality, *e.g.*, higher stability and PLQY, longer PL life times [11,27,55].

Although the NC growth is considered the stage determining the quality of the material and consequently its optical properties, Zhou *et al.* have demonstrated that well-separated and controlled nucleation greatly contributed to the luminescence efficiency of CdTe QDs [56]. According to this method, reorganization of precursors *via* their aging over a period of 9 h at room temperature resulted in the formation of highly crystalline nuclei, which promoted further growth of defect-free particles. Nevertheless, in this case nucleation is also determined by the interaction between precursor species, in particular their electrostatic repulsion leading to decomposition of amorphous nuclei aggregates formed at the very early nucleation step, which in turn is governed by the particle surface.

Obviously, the formation and growth of any kind of NCs, including CdTe, is a multifactorial process, whose analysis requires a complex approach taking into account various aspects such as reactivity of initial precursors, stability of intermediate complexes, electrostatic interactions between monomers and growing species, and their activation energy. The comprehensive description is additionally complicated by changing the role of some certain factors at different stages of particles formation and growth. Moreover, a characterization of the kinetic regimes depends intrinsically on the chemical procedure adopted [57]. Although a creation of a universal theory embracing different sides of the aqueous NCs synthesis seems to be quite difficult, a large number of experimental investigations conducted up to date pave the way for generation of the optimized synthetic protocols for at least the main systems. For instance, basic conditions for the preparation of high quality CdTe NCs capped with TGA, MPA and TG are as follows: a molar ratio of $Cd^{2+}/Te^{2-}$/stabilizer = 1/0.5/1.3, pH 11.5–12, $H_2Te$ gas injection. At the same time, in order to obtain cysteamine-stabilized particles $Cd^{2+}/Te^{2-}$/stabilizer ratio and pH should be adjusted to 1/0.5/2.6 and 6–6.5, respectively.

## 2.4 Postpreparative Treatment and Processing of CdTe Nanocrystals

In certain cases, the quality of the resulting NCs is not satisfactory even by employment of optimal conditions. In order to improve the quality of the as synthesized particles as well as to impart to them new properties and functionalities a postpreparative treatment is applied. Several main methods of postpreparative treatment may be outlined: size selection [1,58], surface modification by means of Cd-thiol complex formation [59,60], photoetching [1,61], ligand exchange [62], ligand modification [63,64]. Since aqueous synthesis usually yields NCs with quite a broad size distribution, as prepared colloids are often subjected to size-selective precipitation consisting in the gradual precipitation of the nanoparticles induced by portion addition of a nonsolvent (*e.g.*, 2-propanol) into a preliminary concentrated NCs solution [1]. Figure 2.7 shows absorption and PL spectra of the size-selected fractions isolated from a crude solution of CdTe NCs. This technique allows the separation of the initial colloid into several fractions of nanoparticles having narrowed size distributions and isolation of the fractions possessing sufficiently enhanced emission compared with ensembles' average level.

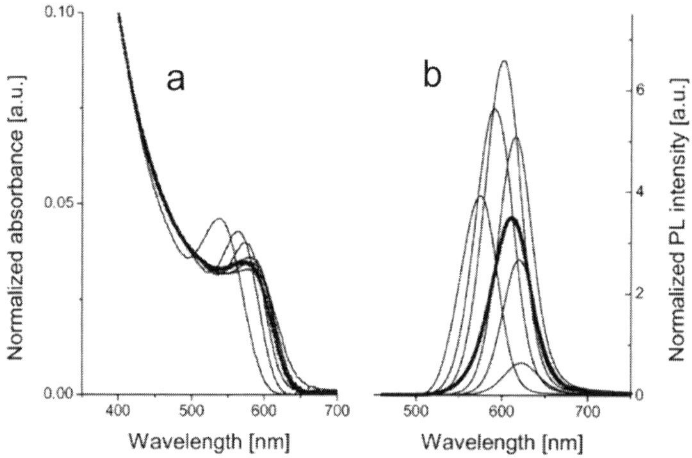


**Figure 2.7** Absorption (a) and PL (b) spectra of size-selected fractions isolated from crude CdTe NCs colloid. Bold lines depict the spectra of as-prepared crude solutions. Reproduced with permission from ref. [1]. Copyright 2002, American Chemical Society.

Recently one more facile separation method was introduced by Ho *et al.*, who applied free-flow electrophoresis for the fractionation of TGA-capped CdTe NCs [58]. Unlike the gel electrophoresis method [65] validated mostly as an analytical tool, this technique is more convenient and preparative. Several variables such as buffer composition, buffer and sample infusion rate, potential field, and channel bed temperature play important roles in successful size separation of nanoparticles. Up to 50% FWHM of the PL peak reduction was achieved, comparing the emission spectra of the separated samples to the original one, as is seen in Fig. 2.8. This technique not only separates the CdTe NC dispersions into more mono-disperse fractions but also can remove unreacted species; it can potentially be a medium for controlled self-organization of nanoparticles under the influence of an electrical field.

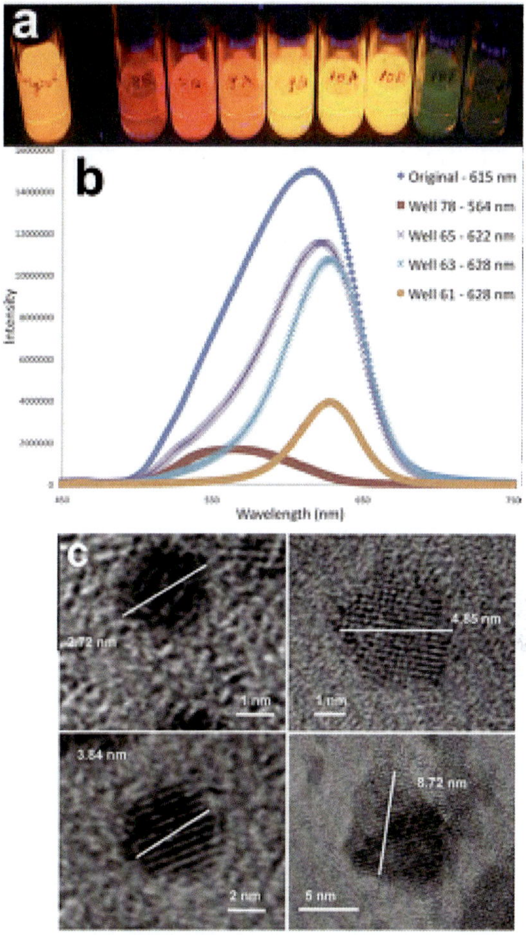


**Figure 2.8** Significant color differences under UV light after free-flow electrophoresis separation (a). Fluorescence spectra of original and separated CdTe NC colloids (b). TEM images of separated CdTe NCs collected from different wells show size difference (c). Adapted with permission from ref. [58]. Copyright 2008, The Royal Society of Chemistry.

The pH-controlled formation of a shell structure of Cd/TGA complexes on the surface of CdTe can also improve the PL QY of the CdTe NCs in aqueous solution [60]. Photochemical etching of CdTe NCs is another powerful tool for their quality enhancement. It has been shown that the photoluminescence efficiency of TGA-capped CdTe colloid increases at the initial stages of photooxidation under irradiation with 400 nm light of a 450 W Xe lamp in oxygen saturated solutions, presumably due to the etching of tellurium trap states, which are responsible for the low fluorescence efficiency [1,66]. In this case, the NCs possessing the most defect states dissolve primarily during photochemical etching. On the contrary, photochemical treatment of CdTe/TGA colloid under inert atmosphere retards its photodegradation and results in a CdTe/CdS core–shell structure formation *via* an intermediate Cd(TGA) complex shell followed by slow generation of sulfide ions reacting with cadmium atoms on CdTe surface [61]. In fact, CdS being a larger band-gap material passivates surface trap states and prevents nonradiative exciton recombination reaching PLQY of 85%. We note that crude aqueous colloids usually contain certain amount of unreacted species, which may contribute to illumination-induced processes. Interestingly, under the same conditions, MPA does not produce sulfide ions, leading to a very limited increment in the PL intensity.

To impart solubility in organic solvents to aqueous particles, strategies for ligand exchange and ligand modification have been developed. Gaponik *et al.* employed a partial exchange of TGA with 1-dodecanethiol (DDT) allowing almost complete (~90%) phase transfer of CdTe NCs synthesized in water to nonpolar organic solvents such as toluene, hexane, chloroform, *etc.* [62]. The optical properties of the transferred QDs exhibit only a moderate alternation in comparison to the particles in water. It was found that acetone plays an important role in an efficient phase transfer of the particles reducing the surface tension at the water/dodecanethiol interface. Utilizing this phase transfer technique, a hybrid aqueous-organic synthesis of core/shell CdTe/ZnS QDs has been developed [30]. This approach affords a facile growth of a ZnS shell in organic solvents at high temperature around CdTe cores originally synthesized in water, to yield small-diameter, highly luminescent material. Combining the treatment with DDT and amphiphilic polymer octadecylamine-modified poly(acrylic acid) (PAA-ODA) in one-pot process, Yuwen *et al.* have developed an interesting method of encapsulation of core/shell MPA-capped aqueous CdTe/CdS QDs into micelles according to the scheme presented in Fig. 2.9 [67]. This polymer coating significantly enhances the photostability and PLQY of the QDs retarding the diffusion of oxygen as well as detachment of thiol ligands, and consequently preventing the oxidation of the particles surface and the release of $Cd^{2+}$ ions into the environment. By controlling the molar ratio of PAA-ODA to QDs, the number of QDs in the micelles can be tuned from a single QD to multiple particles.

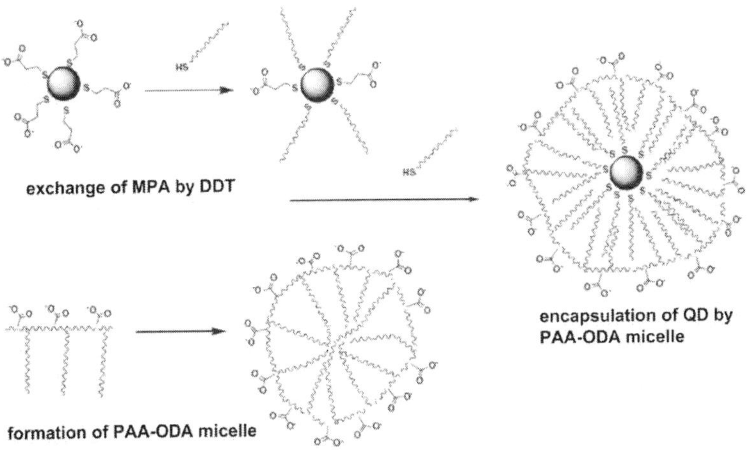


**Figure 2.9** Scheme of the one-pot encapsulation of a QD by amphiphilic PAA-ODA. Reproduced with permission from ref. [67]. Copyright 2011, WILEY VCH Verlag GmbH & Co. KGaA.

Ligand shell modification *via* the interaction between $-COO^-$ groups on the surface of TGA-capped CdTe NCs and $-NH_2$ group of 1-hexadecylamine ensures efficient phase transfer of the QDs from water to chloroform [63]. Adjustment of the pH of the water phase induces the decomposition of $-COO^-Na^+NH_2-$ bonds and results in the reversible transport of the particles back to water. An interesting method of the postpreparative treatment combining application of alkanethiols and alkylammonium surfactant, and allowing controllable bandgap engineering of TGA-capped CdTe NCs was proposed by Akamatsu *et al.* [68]. This technique involves phase transfer of aqueous QDs to toluene employing tetra-n-octylammonium bromide (TOAB) with subsequent injection of various alkanethiols leading to selective dissociation of Te species from crystal surface and simultaneous attachment of alkanethiol at the defect site thus formed. As a result, the PL experienced a significant blue shift whereas the size of particles remained constant.

One more innovative approach of the encapsulation of aqueous CdTe NCs into an amphiphilic polymer have been recently developed in the group of Yang employing a two-step phase transfer of the QDs from water to chloroform by ligand modification with dimethyldioctadecyl-ammonium-bromide (DDAB) and subsequently from chloroform back to water by encapsulation of DDAB-coated particles into amphiphilic poly(ethylene glycol) diglycidyl-grafted poly(maleic anhydride-alt-octadecene), as is illustrated in Fig. 2.10a [69]. The obtained QD-polymer composites exhibited a spherical shape, the average diameter of about 80 nm according to TEM investigation (see Fig. 2.10b) and hydrodynamic diameter of 95 nm measured by dynamic light scattering. The wrapping of CdTe NCs into composite microspheres greatly enhanced their stability under acidic conditions.

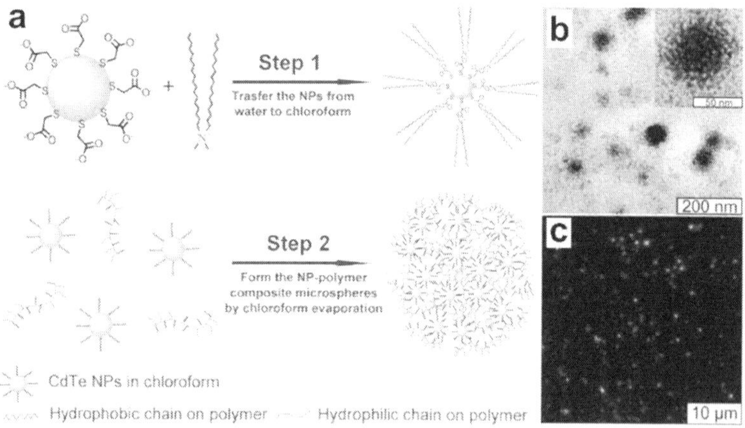


**Figure 2.10** Schematic illustration of the fabrication of water-dispersible CdTe NC–polymer composite microspheres *via* two-step phase transfer (a). TEM images of the microspheres (b). Fluorescence microscope images of the microspheres with orange emission (c). Adapted with permission from ref. [69]. Copyright 2011, The Royal Society of Chemistry.

Direct synthesis of CdTe NCs possessing amphiphilic properties has been recently reported by Dubavik *et al.* [70]. Application of the thiolated poly(ethylene glycol) (mPEG-SH) as a capping ligand enables the synthesis of NCs in water as well as in a variety of organic solvents. This stabilizer facilitates spontaneous direct and complete transfer across phase boundaries without stirring and/or agitation or any additional treatments or medium adjustments, while maintaining the emission properties of the CdTe QDs (see Fig. 2.11).

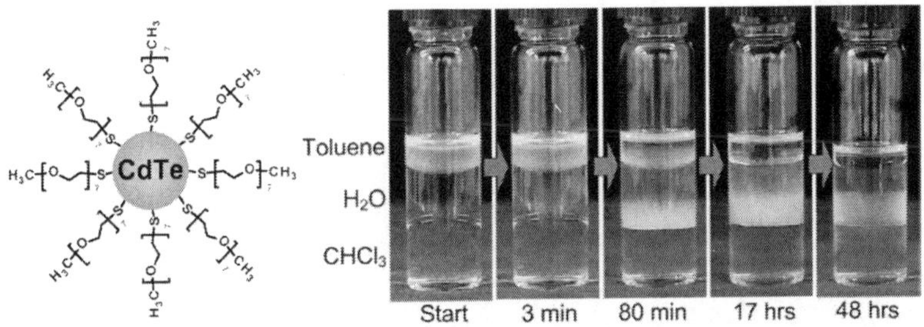


**Figure 2.11** Sketch of a CdTe particle capped by mPEG-SH (left). Photographs of vials demonstrating the spontaneous triphase transfer of CdTe/mPEG-SH NCs from toluene through water into chloroform with time under daylight (right). Adapted with permission from ref. [70]. Copyright 2009, American Chemical Society.

Modification of the thiol-ligand shell with polymerizable-octadecyl-*p*-vinylbenzyldimethylammonium chloride (OVDAC) allows for the efficient transfer of CdTe NCs directly to styrene or styrene/methyl methacrylate with subsequent radical polymerization of the obtained QD–monomer mixture induced by azobisizobutyronitrile (AIBN), yielding transparent highly emitting bulk NC-polymer composites containing up to 10 wt.% of the particles (see Fig. 2.12) [64]. Alternatively, OVDAC-styrene copolymers prepared in a first step can serve as a phase transfer mediator for aqueous CdTe NCs avoiding interaction of the particles with free radicals [71]. Analogously, exploiting copolymer of OVDAC with 9-vinylcarbazole possessing blue fluorescence, in combination with green and red emitting CdTe QDs, Sun *et al.* have demonstrated the fabrication of white LEDs [72]. Application of polymerizable surfactants as both phase transfer agent (didecyl-p-vinylbenzylmethylammonium chloride) and emulsifier (OVDAC) resulted in a preparation of composite CdTe NC-polystyrene microbeads with an average size of ~1.5 μm upon mini-emulsion polymerization [73]. The use of polymerizable surfactants as the emulsifier plays the most important role in effective encapsulation of the aqueous colloidal particles into polymeric beads. At the same time, polymerizable surfactants as the phase transfer agents guarantee homogenous distribution of QDs within the polystyrene matrix and the consequent preservation of the native properties of the NCs.

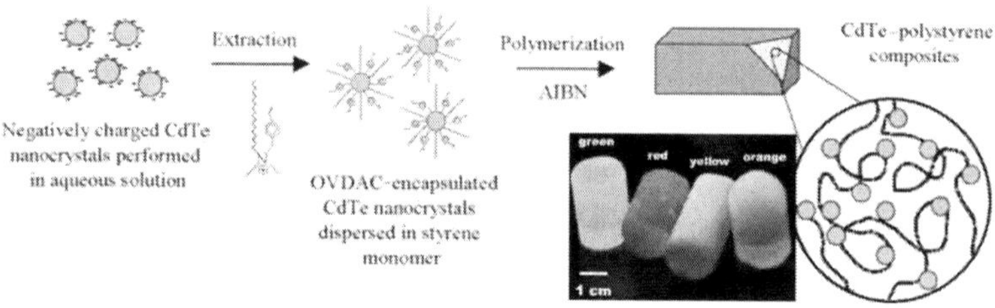


**Figure 2.12** Preparative procedures for the fabrication of fluorescent CdTe-polystyrene composites. Adapted with permission from ref. [64]. Copyright 2003, WILEY VCH Verlag GmbH & Co. KGaA.

Polymers are appropriate candidates to stabilize and protect the NCs, and integrate their functionalities. In addition to polystyrene and polystyrene/polymethyl methacrylate composites, polyurethane has been demonstrated as a solid matrix for encapsulation of CdTe particles [74,75]. Thus, TG-stabilized CdTe NCs dissolved in DMSO formed composites with polyurethane *via* the reaction between –OH groups of TG and –NCO groups of the prepolymer [75]. Water-dispersible polyurethane has been exploited for straightforward fabrication of the composite with cysteine-capped CdTe particles directly in water without any postpreparative treatment [74]. In this case, nanoparticles also act as cross-linking centers reacting with –NCO-terminated prepolymer molecules by amino groups of cysteine, as shown in Fig. 2.13, yielding transparent solid compounds after water evaporation, which retain and even enhance PLQY of the CdTe QDs and impart them photochemical stability.

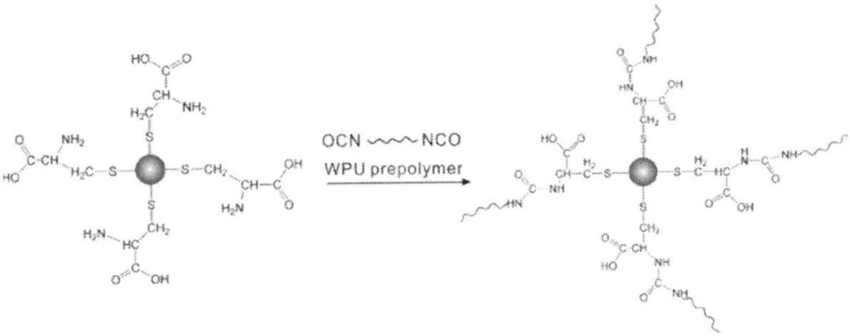


**Figure 2.13** Scheme of the fabrication of the CdTe NC-polyurethane composite. Reproduced with permission from ref. [74]. Copyright 2007, American Chemical Society.

One more material for encapsulation of various nanoparticles is silica [76]. Fabrication of the silica beads incorporating CdTe NCs analogous to those of polystyrene mentioned above has been developed recently by the group of Gao *via* the water-in-oil reverse microemulsion method using cyclohexane, Triton X-100, n-hexanol, and tetraethyl orthosilicate, as a medium, a surfactant, a co-surfactant, and a silica source, respectively [77–79]. The method developed results in the formation of 45–109 nm $SiO_2$ spheres containing in the core one or several CdTe particles, depending on the process conditions (see Fig. 2.14). Additional

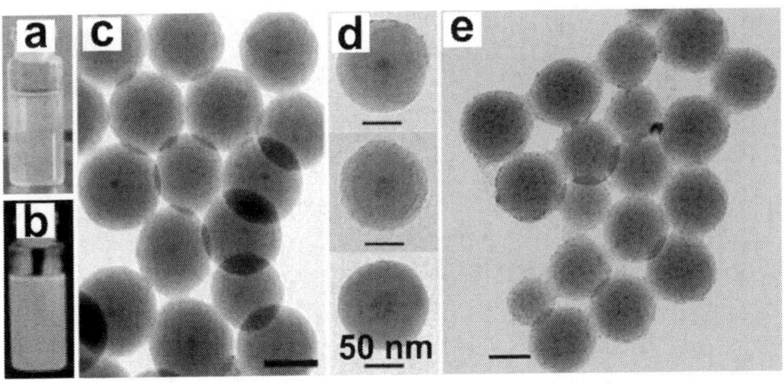


**Figure 2.14** Photographs of an aqueous dispersion of CdTe@$SiO_2$ particles under daylight (a) and UV light (b). Silica beads with different distribution of the CdTe crystals: one QD located in the center (c), increasing number of the QDs in the center (d, from top to bottom), QDs evenly distributed over the whole beads (e). Adapted with permission from ref. [78]. Copyright 2005, WILEY VCH Verlag GmbH & Co. KGaA.

incubation of the as-prepared CdTe colloid in ammonia solution enhances the PLQY of the QDs and moreover retains excellent emission with the efficiency of up to 47% for the resulting CdTe@$SiO_2$ core/shell particles [77]. A silica shell impedes the diffusion of oxygen from the surrounding environment preventing CdTe NCs from photo-oxidation and consequently blocks diffusion of Cd ions reducing cytotoxicity of such core/shell particles making them promising material for bio-labeling. Further surface modification of the silica coated QDs with amino groups provides a route for covalent linking of biomolecules [79].

## 2.5 CdTe-Based Core/Shell Nanocrystals

Microwave irradiation and ultrasonic treatment as well as successive ion-layer adsorption and reaction (SILAR) have been declared as efficient techniques for the preparation of core/shell CdTe based particles. Thus, using microwave irradiation the first synthesis of core/shell/shell MPA-capped CdTe/CdS/ZnS NCs directly in water was recently introduced [29]. This protocol comprises three steps: CdTe core preparation, coating the core by CdS shell, and a second shell (ZnS) formation (see Fig. 2.15), in order to eliminate possible lattice strain induced by a large lattice mismatch between CdTe core and the outer ZnS shell. NCs after each step undergo purification by re-precipitation, which is necessary for separation of unreacted species. The resulting QDs were found to be very stable in biological media and non-cytotoxic, that makes them very promising candidates for biolabeling.

Another material for bioimaging application—type II CdTe/CdSe core/shell QDs with NIR fluorescence—was prepared applying the SILAR technique [80]. According to this recipe, L-cysteine-stabilized CdTe core was covered with 1–6 CdSe layers by sequential injection of $Cd^{2+}$/cysteine and KHSe solutions upon reflux, which resulted in the preparation of NCs emitting in the red–NIR region of 600–850 nm with the highest QY = 12%. The SILAR technique ensures very efficient control of shell thickness tuning the behavior of core/shell CdTe/CdS QDs from type I to type II structure by gradual deposition of several shell layers, indicated by a change of their optical properties including sufficient enhancement of QY and PL life time compared with those of initial core particles [81]. In addition to microwave irradiation, ultrasonic treatment has been applied for CdS shell formation on CdTe NCs [26]. In this case, acoustic cavitation facilitates the decomposition of thiourea, which releases $S^{2-}$ reacting with $Cd^{2+}$ present in solution with subsequent creation of a gradient shell.

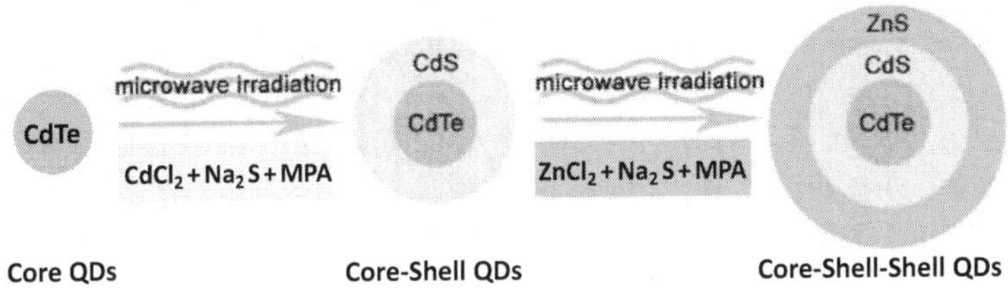


**Figure 2.15** Schematic illustration of the microwave-assisted synthesis of aqueous CdTe/CdS/ZnS core/shell/shell QDs. Reproduced with permission from ref. [29]. Copyright 2009, WILEY VCH Verlag GmbH & Co. KGaA.

We have to admit, unfortunately, some researchers often do not satisfactorily clearly demonstrate the formation of core/shell structures in their reports. In our opinion, some reports, especially those claiming a one-pot aqueous synthesis for core/shell NCs, when the precursors for both the core and the shell materials are added and react simultaneously, give rise to realistic criticism, since in this case the formation of an alloyed structure rather than a classical core/shell is much more probable.

## 2.6 CdTe-Based Alloyed Nanocrystals

All the synthetic approaches described above ensure the synthesis of CdTe NCs emitting mainly in the visible region. To extend their emission to longer wavelengths, one uses alloying of the NCs with materials possessing narrower bandgaps and quite similar lattice parameters to CdTe. In the case of aqueous CdTe, approaches for alloying *via* partial replacement of both the cation and the anion have been developed.

### 2.6.1 CdHgTe

An appropriate candidate for CdTe cation alloying is HgTe, which has a very close lattice constant (cf. 6.477 and 6.462 Å for bulk zinc blende CdTe and HgTe, respectively [82]). The aqueous synthesis of NIR emitting alloyed CdHgTe and core/shell CdHgTe/CdS has been first reported by Harrison, Rogach, and others [23,83,84]. This approach based on the incorporation of $Hg^{2+}$ ions into already prepared CdTe NCs, stabilized by 2-mercaptoethanol or TG, involves several steps: the synthesis of the initial NCs followed by the subsequent formation of a HgTe shell. To shift the PL to wavelengths longer than 1000 nm, this layer-by-layer procedure has to include the creation of 2–3 HgTe layers separated by CdTe layers upon addition of different amounts of $Hg^{2+}$ ions to the CdTe colloidal solutions corresponding to 20%, 40%, 80% and 120% of the cadmium content of the aliquots, as is schematically shown in Fig. 2.16 [83].

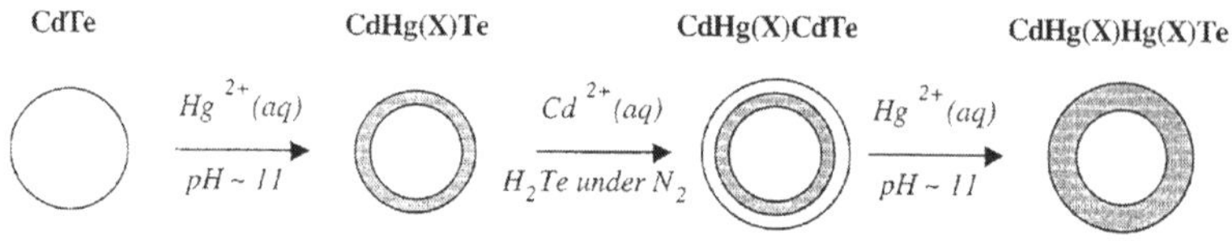


**Figure 2.16** Scheme of the synthetic route for obtaining $Cd_xHg_{1-x}Te$ NCs. Reproduced with permission from ref. [83]. Copyright 2000, Elsevier.

A similar approach was applied for the preparation of CdHgTe nanorods using CdTe nanorods as a template [33]. The authors have assumed that the redistribution of mercury and cadmium ions in the growing QDs most probably leads to the formation of $Cd_xHg_{1-x}Te$ alloys rather than to discretely layered NCs. The shell formation is more obvious in the hybrid approach, when CdTe or CdHgTe particles were transferred from water to organic with subsequent coating by ZnS shell employing an organometallic route yielding small-diameter, highly NIR-luminescent NCs [30]. We note that water strongly absorbs in the NIR and IR spectral region starting from approximately 1150 nm (its transmission measured in 1 cm cuvette at this wavelength is ca. 35%; see Fig. 2.17a) sufficiently obscuring the luminescence of samples [22]. Therefore, solvents with lower absorption have to be used for PL spectra acquiring ($D_2O$, transfer to organics) or measurements from dried NCs deposited onto a transparent wafer.

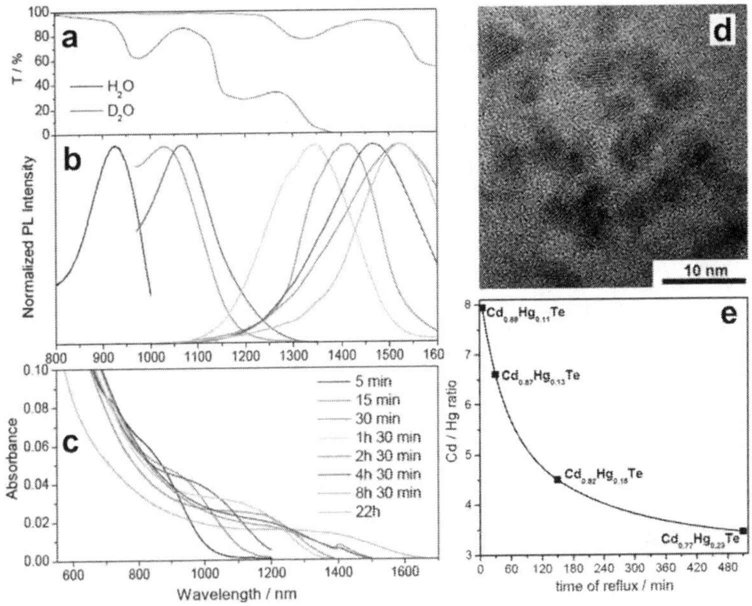


**Figure 2.17** Transmission spectra of $H_2O$ and $D_2O$ (a), evolution of the PL (b) and absorption spectra (c) of $Cd_{0.95}Hg_{0.05}Te$ NCs as a function of the reflux time, measured in $D_2O$. High-resolution TEM image of $Cd_{0.95}Hg_{0.05}Te$/TGA NCs (d). Cd/Hg ratio and corresponding composition of the TGA-capped $Cd_{1-x}Hg_xTe$ particles obtained from the initial feed ratio $Cd^{2+}/Hg^{2+}$ = 95/5 as a function of the reflux time (e). Adapted with permission from ref. [22]. Copyright 2009, The Royal Society of Chemistry.

The incorporation method is limited in terms of the achievable spectral position of the PL maximum, *i.e.*, the tuneability of the PL band. The ultimate value reachable with this synthesis is 1100 nm for the PL maximum and 40% for the PLQY of the corresponding NCs. The newly developed facile one-pot synthesis provides a wide range of alloyed TGA stabilized $Cd_xHg_{1-x}Te$ NC batches possessing a strong PL within the spectral range of 640 to 1600 nm (see Fig. 2.17b) [22]. The recipe is very similar to the synthesis of pure CdTe and includes partial replacement of the Cd salt by the Hg precursor. In contrast to the incorporation technique, the nucleation and the growth of alloyed NCs using the addition of only a small amount of the mercury precursor leads to a significant decrease of the bandgap and correspondingly to a shift of the PL to longer wavelengths. The authors have suggested that the synthesis of the $Cd_xHg_{1-x}Te$ particles follows approximately the same NC growth kinetics as the synthesis of undoped TGA-capped CdTe. Thus, the shift of the first excitonic absorption maxima and the PL maxima during the synthesis of the alloyed NCs occurs only due to the incorporation of the narrower bandgap material, that was confirmed by an elemental analysis (see Fig. 2.17e). At the beginning of the growth the Cd content in the NCs is less than in the initial feed ratio (Cd/Hg = 95/5) and continuously decreases due to the approximately 20 times lower solubility of HgTe compared with CdTe in water. A similar relationship was observed by Sun *et al.*, having obtained a composition of $Cd_{0.23}Hg_{0.77}Te$ from the initial feed ratio of Cd/Hg = 1/1 [24]. Here we would like the reader to note the necessity of reconsideration of PLQYs of those NIR-emitting samples measured using as a standard dye IR-26. Its quantum efficiency has been re-examined lately and revealed as 0.05% (at least in 1,2-dichloroethane) that is an order of magnitude lower than previously reported value [85]. Therefore, PLQYs may have been previously overestimated by a factor of ~10.

### 2.6.2 CdSeTe

Alloying *via* replacement of Te ions in CdTe crystal lattice by another anion is additional possibility for the bandgap engineering. Analogous to the CdHgTe NCs preparation, the incorporation technique has been applied for a synthesis of alloyed L-cysteine capped CdSeTe and CdSeTe/ZnS particles using CdTe as initial material [86,87]. Although CdSe possesses a wider bandgap than CdTe (1.74 eV for wurtzite structure [82]), the alloyed QDs demonstrate nonlinear dependence of the bandgap on their composition, denoted as optical bowing effect decreasing the optical bandgap of alloyed particles compared with CdTe NCs of the same size and enabling a shift of the PL up to 814 nm [86]. A ZnS shell coating ensures high PLQY = 70% of resulted core/shell particles [87]. Piven *et al.* developed an one-pot synthesis of alloyed TGA-capped CdSeTe and CdSTe NCs using simultaneous injection of NaHSe ($Na_2S$) and NaHTe into the Cd precursor solution [21].

In this case the emission efficiency was approximately one order of magnitude lower than that for thiol-capped CdTe NCs being in the range of 2–4%, which is determined by different energetics of thiol-related trap surface states for CdTe and CdSe serving as nonradiative channels for CdSe NCs [88]. The measured values of Stokes shifts for different compositions of $CdSe_xTe_{1-x}$ NCs are comparable or even smaller than those for the reference CdTe NCs, indicating that the PL of alloys is mainly determined by the exciton recombination rather than by the surface trap emission. The above-mentioned results suggest that L-cysteine is a beneficial stabilizer for alloyed CdSeTe QDs.

## 2.7 Summary and Outlook

As is clearly seen from the work reviewed in this chapter, colossal progress has been achieved during the last decade in the colloidal aqueous synthesis of CdTe NCs, including both its practical preparative aspect and theoretical insight into the processes and mechanisms responsible for the formation of nanoparticles. As a result of intensive investigations many straightforward synthetic techniques providing high quality materials have been developed. Thereby prepared CdTe QDs demonstrate excellent emission properties; owing to their size control, they cover a wide visible spectral region from the green to the NIR, while alloying of CdTe *via* partial cation or anion replacement leads to further shift of the emission to the longer wavelengths up to 1600 nm. Recently demonstrated synthesis of core/shell and core/shell/shell CdTe based structures is a great step forward, delivering colloidally and chemically stable materials especially promising for applications requiring continuous contact with oxidative environment, *e.g.*, *in vivo* bioimaging. Efficient strategies for the encapsulation of CdTe particles into polymeric micelles, silica spheres also provide robust nanoscale materials preserving and even enhancing the optical properties of the parental NCs. Obtaining amphiphilic NCs and their composites possessing versatile solubility and thus being potentially capable of penetrating through cell membranes is an additional advantage of aqueous synthesis.

A vast variability of the functionalization of the CdTe NCs achievable both directly in the synthesis during their growth and by means of the postpreparative treatment is a key factor for their processing and subsequent application in various fields from life sciences to photonics. Aqueous CdTe NCs being functionalized directly in the synthesis *via* application of various thiol stabilizers are off-the-shelf building blocks for the fabrication of different hybrid structures utilizing assembly approaches. Assembly of nanoparticles into superstructures of greater complexity is a necessary step in the utilization of nanomaterials, which enhances the spectrum of their properties. It is believed that the complexity of nanoparticulate architectures will increase not only with respect to spatial dimensions but also along the temporal coordinate, leading to dynamic nanometer-scale systems [89]. Additional dimensions for the control of the CdTe NCs properties are provided by their postpreparative treatment employing innovative strategies for their further functionalization and processing, which pave the ways for the creation of novel complex materials, *e.g.*, polymer composites.

## Abbreviations

| | |
|---|---|
| AIBN | 2,2′-azobisizobutyronitrile |
| DDAB | dimethyldioctadecylammonium bromide |
| DDT | 1-dodecanethiol |
| FWHM | full width at half-maximum |
| LED | light-emitting diode |
| MPA | 3-mercaptopropionic acid |
| mPEG-SH | thiolated poly(ethylene glycol) |
| NC | nanocrystal |
| NIR | near infrared |
| OA | oriented attachment |
| OR | Ostwald ripening |
| OVDAC | octadecyl-*p*-vinylbenzyldimethylammonium chloride |
| PAA-ODA | octadecylamine-modified poly(acrylic acid) |
| PL | photoluminescence |
| QD | quantum dot |
| QY | quantum yield |
| SILAR | successive ion-layer adsorption and reaction |
| (HR)TEM | (high-resolution) transmission electron microscopy |
| TG | 3-mercapto-l,2-propanediol (thioglycerol) |
| TGA | thioglycolic acid |
| TOAB | tetra-n-octylammonium bromide |

## References

1. Gaponik, N., Talapin, D. V., Rogach, A. L., Hoppe, K., Shevchenko, E. V., Kornowski, A., Eychmüller, A., and Weller, H. (2002). Thiol-capping of CdTe nanocrystals: an alternative to organometallic synthetic routes, *J. Phys. Chem. B*, **106**, pp. 7177–7185.
2. Gaponik, N., and Rogach, A. L. (2010). Thiol-capped CdTe nanocrystals: progress and perspectives of the related research fields, *Phys. Chem. Chem. Phys.*, **12**, pp. 8685–8693.
3. Henglein, A. (1989). Small-particle research—physicochemical properties of extremely small colloidal metal and semiconductor particles, *Chem. Rev.*, **89**, pp. 1861–1873.
4. Eychmüller, A. (2004). *Nanoparticles: From Theory to Application*, ed. G. Schmid, Chapter 3.1.1 "Synthesis and Characterization of II–VI Nanoparticles" (WILEY-VCH Verlag GmbH & Co. KGaA, Weinheim) pp. 50–79.
5. Gaponik, N., Hickey, S. G., Dorfs, D., Rogach, A. L., and Eychmüller, A. (2010). Progress in the light emission of colloidal semiconductor nanocrystals, *Small*, **6**, pp. 1364–1378.
6. Gaponik, N., and Rogach, A. L. (2008). *Semiconductor Nanocrystal Quantum Dots: Synthesis, Assembly, Spectroscopy and Applications*, ed. A. L. Rogach, Chapter "Aqueous synthesis of semiconductor nanocrystals" (Springer, Wien, New York) pp. 73–100.
7. Li, Y., Jing, L., Qiao, R., and Gao, M. (2011). Aqueous synthesis of CdTe nanocrystals: progresses and perspectives, *Chem. Commun.*, **47**, pp. 9293–9311.
8. Resch, U., Weller, H., and Henglein, A. (1989). Photochemistry and radiation chemistry of colloidal semiconductors. 33. Chemical changes and fluorescence in CdTe and ZnTe, *Langmuir*, **5**, pp. 1015–1020.
9. Rajh, T., Micic, O. I., and Nozik, A. J. (1993). Synthesis and characterization of surface-modified colloidal cadmium telluride quantum dots, *J. Phys. Chem.*, **97**, pp. 11999–12003.
10. Rogach, A. L., Katsikas, L., Kornowski, A., Su, D., Eychmüller, A., and Weller, H. (1996). Synthesis and characterization of thiol-stabilized CdTe nanocrystals, *Ber. Bunsenges. Phys. Chem.*, **100**, pp. 1772–1778.
11. Rogach, A. L., Franzl, T., Klar, T. A., Feldmann, J., Gaponik, N., Lesnyak, V., Shavel, A., Eychmüller, A., Rakovich, Y. P., and Donegan, J. F. (2007). Aqueous synthesis of thiol-capped CdTe nanocrystals: state-of-the-art, *J. Phys. Chem. C*, **111**, pp. 14628–14637.

12. He, Y., Sai, L. M., Lu, H. T., Hu, M., Lai, W. Y., Fan, Q. L., Wang, L. H., and Huang, W. (2007). Microwave-assisted synthesis of water-dispersed CdTe nanocrystals with high luminescent efficiency and narrow size distribution, *Chem. Mater.*, **19**, pp. 359–365.

13. Grabolle, M., Spieles, M., Lesnyak, V., Gaponik, N., Eychmüller, A., and Resch-Genger, U. (2009). Determination of the fluorescence quantum yield of quantum dots: suitable procedures and achievable uncertainties, *Anal. Chem.*, **81**, pp. 6285–6294.

14. Bendall, J. S., Paderi, M., Ghigliotti, F., Li Pira, N., Lambertini, V., Lesnyak, V., Gaponik, N., Visimberga, G., Eychmüller, A., Torres, C. M. S., Welland, M. E., Gieck, C., and Marchese, L. (2010). Layer-by-layer all-inorganic quantum-dot-based LEDs: a simple procedure with robust performance, *Adv. Funct. Mater.*, **20**, pp. 3298–3302.

15. Rogach, A. L., Gaponik, N., Lupton, J. M., Bertoni, C., Gallardo, D. E., Dunn, S., Li Pira, N., Paderi, M., Repetto, P., Romanov, S. G., O'Dwyer, C., Sotomayor Torres, C. M., and Eychmüller, A. (2008). Light-emitting diodes with semiconductor nanocrystals, *Angew. Chem. Int. Ed.*, **47**, pp. 6538–6549.

16. Cicek, N., Nizamoglu, S., Ozel, T., Mutlugun, E., Karatay, D. U., Lesnyak, V., Otto, T., Gaponik, N., Eychmüller, A., and Demir, H. V. (2009). Structural tuning of color chromaticity through nonradiative energy transfer by interspacing CdTe nanocrystal monolayers, *Appl. Phys. Lett.*, **94**, pp. 061105–061103.

17. Franzl, T., Klar, T. A., Schietinger, S., Rogach, A. L., and Feldmann, J. (2004). Exciton recycling in graded gap nanocrystal structures, *Nano Lett.*, **4**, pp. 1599–1603.

18. Rakovich, A., Sukhanova, A., Bouchonville, N., Lukashev, E., Oleinikov, V., Artemyev, M., Lesnyak, V., Gaponik, N., Molinari, M., Troyon, M., Rakovich, Y. P., Donegan, J. F., and Nabiev, I. (2010). Resonance energy transfer improves the biological function of bacteriorhodopsin within a hybrid material built from purple membranes and semiconductor quantum dots, *Nano Lett.*, **10**, pp. 2640–2648.

19. Susha, A. S., Javier, A. M., Parak, W. J., and Rogach, A. L. (2006). Luminescent CdTe nanocrystals as ion probes and pH sensors in aqueous solutions, *Colloids Surf. A*, **281**, pp. 40–43.

20. Nabiev, I., Mitchell, S., Davies, A., Williams, Y., Kelleher, D., Moore, R., Gun'ko, Y. K., Byrne, S., Rakovich, Y. P., Donegan, J. F., Sukhanova, A., Conroy, J., Cottell, D., Gaponik, N., Rogach, A., and Volkov, Y. (2007). Nonfunctionalized nanocrystals can exploit a cell's active transport

machinery delivering them to specific nuclear and cytoplasmic compartments, *Nano Lett.*, **7**, pp. 3452–3461.

21. Piven, N., Susha, A. S., Döblinger, M., and Rogach, A. L. (2008). Aqueous synthesis of alloyed $CdSe_xTe_{1-x}$ nanocrystals, *J. Phys. Chem. C*, **112**, pp. 15253–15259.

22. Lesnyak, V., Lutich, A., Gaponik, N., Grabolle, M., Plotnikov, A., Resch-Genger, U., and Eychmüller, A. (2009). One-pot aqueous synthesis of high quality near infrared emitting $Cd_{1-x}Hg_xTe$ nanocrystals, *J. Mater. Chem.*, **19**, pp. 9147–9152.

23. Rogach, A. L., Harrison, M. T., Kershaw, S. V., Kornowski, A., Burt, M. G., Eychmüller, A., and Weller, H. (2001). Colloidally prepared CdHgTe and HgTe quantum dots with strong near-infrared luminescence, *Phys. Status Solidi B*, **224**, pp. 153–158.

24. Sun, H. Z., Zhang, H., Ju, J., Zhang, J. H., Qian, G., Wang, C. L., Yang, B., and Wang, Z. Y. (2008). One-step synthesis of high-quality gradient CdHgTe nanocrystals: a prerequisite to prepare CdHgTe-polymer bulk composites with intense near-infrared photoluminescence, *Chem. Mater.*, **20**, pp. 6764–6769.

25. He, Y., Lu, H.-T., Sai, L.-M., Lai, W.-Y., Fan, Q.-L., Wang, L.-H., and Huang, W. (2006). Microwave-assisted growth and characterization of water-dispersed CdTe/CdS core–shell nanocrystals with high photoluminescence, *J. Phys. Chem. B*, **110**, pp. 13370–13374.

26. Wang, C., Zhang, H., Zhang, J., Li, M., Sun, H., and Yang, B. (2007). Application of ultrasonic irradiation in aqueous synthesis of highly fluorescent CdTe/CdS core–shell nanocrystals, *J. Phys. Chem. C*, **111**, pp. 2465–2469.

27. Zhao, D., He, Z., Chan, W. H., and Choi, M. M. F. (2008). Synthesis and characterization of high-quality water-soluble near-infrared-emitting CdTe/CdS quantum dots capped by N-acetyl-l-cysteine *via* hydrothermal method, *J. Phys. Chem. C*, **113**, pp. 1293–1300.

28. Law, W. C., Yong, K. T., Roy, I., Ding, H., Hu, R., Zhao, W., and Prasad, P. N. (2009). Aqueous-phase synthesis of highly luminescent CdTe/ZnTe core/shell quantum dots optimized for targeted bioimaging, *Small*, **5**, pp. 1302–1310.

29. He, Y., Lu, H. T., Sai, L. M., Su, Y. Y., Hu, M., Fan, C. H., Huang, W., and Wang, L. H. (2008). Microwave synthesis of water-dispersed CdTe/CdS/ZnS core–shell–shell quantum dots with excellent photostability and biocompatibility, *Adv. Mater.*, **20**, pp. 3416–3421.

30. Tsay, J. M., Pflughoefft, M., Bentolila, L. A., and Weiss, S. (2004). Hybrid approach to the synthesis of highly luminescent CdTe/ZnS and CdHgTe/ZnS nanocrystals, *J. Am. Chem. Soc.*, **126**, pp. 1926–1927.
31. Li, J., Hong, X., Li, D., Zhao, K., Wang, L., Wang, H., Du, Z., Li, J., Bai, Y., and Li, T. (2004). Mixed ligand system of cysteine and thioglycolic acid assisting in the synthesis of highly luminescent water-soluble CdTe nanorods, *Chem. Commun.*, pp. 1740–1741.
32. Zhang, H., Wang, D., and Möhwald, H. (2006). Ligand-selective aqueous synthesis of one-dimensional CdTe nanostructures, *Angew. Chem. Int. Ed.*, **45**, pp. 748–751.
33. Tang, B., Yang, F., Lin, Y., Zhuo, L., Ge, J., and Cao, L. (2007). Synthesis and characterization of wavelength-tunable, water-soluble, and near-infrared-emitting CdHgTe nanorods, *Chem. Mater.*, **19**, pp. 1212–1214.
34. Smith, A. M., and Nie, S. (2010). Semiconductor nanocrystals: structure, properties, and band gap engineering, *Acc. Chem. Res.*, **43**, pp. 190–200.
35. Tang, Z., Kotov, N. A., and Giersig, M. (2002). Spontaneous organization of single CdTe nanoparticles into luminescent nanowires, *Science*, **297**, pp. 237–240.
36. Volkov, Y., Mitchell, S., Gaponik, N., Rakovich, Y. P., Donegan, J. F., Kelleher, D., and Rogach, A. L. (2004). *In-situ* observation of nanowire growth from luminescent CdTe nanocrystals in a phosphate buffer solution, *Chem. Phys. Chem.*, **5**, pp. 1600–1602.
37. Srivastava, S., Santos, A., Critchley, K., Kim, K.-S., Podsiadlo, P., Sun, K., Lee, J., Xu, C., Lilly, G. D., Glotzer, S. C., and Kotov, N. A. (2010). Light-controlled self-assembly of semiconductor nanoparticles into twisted ribbons, *Science*, **327**, pp. 1355–1359.
38. Tang, Z., Zhang, Z., Wang, Y., Glotzer, S. C., and Kotov, N. A. (2006). Self-assembly of CdTe nanocrystals into free-floating sheets, *Science*, **314**, pp. 274–278.
39. Gaponik, N., Wolf, A., Marx, R., Lesnyak, V., Schilling, K., and Eychmüller, A. (2008). Three-dimensional self-assembly of thiol-capped CdTe nanocrystals: gels and aerogels as building blocks for nanotechnology, *Adv. Mater.*, **20**, pp. 4257–4262.
40. Lesnyak, V., Wolf, A., Dubavik, A., Borchardt, L., Voitekhovich, S. V., Gaponik, N., Kaskel, S., and Eychmüller, A. (2011). 3D assembly of semiconductor and metal nanocrystals: hybrid CdTe/Au structures with controlled content, *J. Am. Chem. Soc.*, **133**, pp. 13413–13420.

41. He, Y., Lu, H. T., Sai, L. M., Lai, W. Y., Fan, Q. L., Wang, L. H., and Huang, W. (2006). Synthesis of CdTe nanocrystals through program process of microwave irradiation, *J. Phys. Chem. B*, **110**, pp. 13352–13356.

42. Li, L., Qian, H., and Ren, J. (2005). Rapid synthesis of highly luminescent CdTe nanocrystals in the aqueous phase by microwave irradiation with controllable temperature, *Chem. Commun.*, pp. 528–530.

43. Nakashima, T., Kobayashi, Y., and Kawai, T. (2009). Optical activity and chiral memory of thiol-capped CdTe nanocrystals, *J. Am. Chem. Soc.*, **131**, pp. 10342–10343.

44. Zhou, Y., Yang, M., Sun, K., Tang, Z., and Kotov, N. A. (2010). Similar topological origin of chiral centers in organic and nanoscale inorganic structures: effect of stabilizer chirality on optical isomerism and growth of CdTe nanocrystals, *J. Am. Chem. Soc.*, **132**, pp. 6006–6013.

45. Zou, L., Gu, Z., Zhang, N., Zhang, Y., Fang, Z., Zhu, W., and Zhong, X. (2008). Ultrafast synthesis of highly luminescent green- to near infrared-emitting CdTe nanocrystals in aqueous phase, *J. Mater. Chem.*, **18**, pp. 2807–2815.

46. Boldt, K., Bruns, O. T., Gaponik, N., and Eychmüller, A. (2006). Comparative examination of the stability of semiconductor quantum dots in various biochemical buffers, *J. Phys. Chem. B*, **110**, pp. 1959–1963.

47. Zhang, H., Wang, D., Yang, B., and Möhwald, H. (2006). Manipulation of aqueous growth of CdTe nanocrystals to fabricate colloidally stable one-dimensional nanostructures, *J. Am. Chem. Soc.*, **128**, pp. 10171–10180.

48. Yin, S., Huang, F., Zhang, J., Zheng, J., and Lin, Z. (2011). The effects of particle concentration and surface charge on the oriented attachment growth kinetics of CdTe nanocrystals in $H_2O$, *J. Phys. Chem. C*, **115**, pp. 10357–10364.

49. Zhang, H., Liu, Y., Wang, C., Zhang, J., Sun, H., Li, M., and Yang, B. (2008). Directing the growth of semiconductor nanocrystals in aqueous solution: role of electrostatics, *Chem. Phys. Chem.*, **9**, pp. 1309–1316.

50. Zhang, H., Liu, Y., Zhang, J., Wang, C., Li, M., and Yang, B. (2008). Influence of interparticle electrostatic repulsion in the initial stage of aqueous semiconductor nanocrystal growth, *J. Phys. Chem. C*, **112**, pp. 1885–1889.

51. Shavel, A., Gaponik, N., and Eychmüller, A. (2006). Factors governing the quality of aqueous CdTe nanocrystals: calculations and experiment, *J. Phys. Chem. B*, **110**, pp. 19280–19284.

52. Guo, J., Yang, W. L., and Wang, C. C. (2005). Systematic study of the photoluminescence dependence of thiol-capped CdTe nanocrystals on the reaction conditions, *J. Phys. Chem. B*, **109**, pp. 17467–17473.

53. Wang, C. L., Zhang, H., Xu, S. H., Lv, N., Liu, Y., Li, M. J., Sun, H. Z., Zhang, J. H., and Yang, B. (2009). Sodium-citrate-assisted synthesis of aqueous CdTe nanocrystals: giving new insight into the effect of ligand shell, *J. Phys. Chem. C*, **113**, pp. 827–833.

54. Rogach, A. L. (2000). Nanocrystalline CdTe and CdTe(S) particles: wet chemical preparation, size-dependent optical properties and perspectives of optoelectronic applications, *Mater. Sci. Eng. B*, **69–70**, pp. 435–440.

55. Zhao, K., Li, J., Wang, H. Z., Zhuang, J. Q., and Yang, W. S. (2007). Stoichiometric ratio dependent photoluminescence quantum yields of the thiol capping CdTe nanocrystals, *J. Phys. Chem. C*, **111**, pp. 5618–5621.

56. Zhou, D., Han, J., Liu, Y., Liu, M., Zhang, X., Zhang, H., and Yang, B. (2010). Nucleation of aqueous semiconductor nanocrystals: a neglected factor for determining the photoluminescence, *J. Phys. Chem. C*, **114**, pp. 22487–22492.

57. Ferreira, D. L., Silva, F. O., Viol, L. C. D. S., Licinio, P., Valadares, M., Cury, L. A., Schiavon, M. A., and Alves, J. L. A. (2009). Growth kinetics of CdTe colloidal nanocrystals, *J. Chem. Phys.*, **131**, pp. 084712.

58. Ho, S., Critchley, K., Lilly, G. D., Shim, B., and Kotov, N. A. (2009). Free flow electrophoresis for the separation of CdTe nanoparticles, *J. Mater. Chem.*, **19**, pp. 1390–1394.

59. Gao, M., Kirstein, S., Möhwald, H., Rogach, A. L., Kornowski, A., Eychmüller, A., and Weller, H. (1998). Strongly photoluminescent CdTe nanocrystals by proper surface modification, *J. Phys. Chem. B*, **102**, pp. 8360–8363.

60. Zhang, H., Zhou, Z., Yang, B., and Gao, M. (2003). The influence of carboxyl groups on the photoluminescence of mercaptocarboxylic acid-stabilized CdTe nanoparticles, *J. Phys. Chem. B*, **107**, pp. 8–13.

61. Bao, H., Gong, Y., Li, Z., and Gao, M. (2004). Enhancement effect of illumination on the photoluminescence of water-soluble CdTe nanocrystals: toward highly fluorescent CdTe/CdS core–shell structure, *Chem. Mater.*, **16**, pp. 3853–3859.

62. Gaponik N., Talapin D. V., Rogach A. L., Eychmüller A., and Weller H. (2002). Efficient phase transfer of luminescent thiol-capped nanocrystals: from water to nonpolar organic solvents, *Nano Lett.*, **2**, pp. 803–806.

63. Jiang, H., and Jia, J. (2008). Complete reversible phase transfer of luminescent CdTe nanocrystals mediated by hexadecylamine, *J. Mater. Chem.*, **18**, pp. 344–349.

64. Zhang, H., Cui, Z., Wang, Y., Zhang, K., Ji, X., Lu, C., Yang, B., and Gao, M. (2003). From water-soluble CdTe nanocrystals to fluorescent nanocrystal-polymer transparent composites using polymerizable surfactants, *Adv. Mater.*, **15**, pp. 777–780.

65. Eychmüller, A., Katsikas, L., and Weller, H. (1990). Photochemistry of semiconductor colloids. 35. Size separation of colloidal cadmium sulfide by gel electrophoresis, *Langmuir*, **6**, pp. 1605–1608.

66. Borchert, H., Talapin, D. V., Gaponik, N., McGinley, C., Adam, S., Lobo, A., Möller, T., and Weller, H. (2003). Relations between the photoluminescence efficiency of CdTe nanocrystals and their surface properties revealed by synchrotron XPS, *J. Phys. Chem. B*, **107**, pp. 9662–9668.

67. Yuwen, L., Bao, B., Liu, G., Tian, J., Lu, H., Luo, Z., Zhu, X., Boey, F., Zhang, H., and Wang, L. (2011). One-pot encapsulation of luminescent quantum dots synthesized in aqueous solution by amphiphilic polymers, *Small*, **7**, pp. 1456–1463.

68. Akamatsu, K., Tsuruoka, T., and Nawafune, H. (2005). Band gap engineering of CdTe nanocrystals through chemical surface modification, *J. Am. Chem. Soc.*, **127**, pp. 1634–1635.

69. Ning, Y., Zhang, H., Han, J., Yang, C., Liu, Y., Zhou, D., and Yang, B. (2011). Versatile fabrication of water-dispersible nanoparticle-amphiphilic copolymer composite microspheres with specific functionalities, *J. Mater. Chem.*, **21**, pp. 6837–6843.

70. Dubavik, A., Lesnyak, V., Thiessen, W., Gaponik, N., Wolff, T., and Eychmüller, A. (2009). Synthesis of amphiphilic CdTe nanocrystals, *J. Phys. Chem. C*, **113**, pp. 4748–4750.

71. Zhang, H., Wang, C., Li, M., Ji, X., Zhang, J., and Yang, B. (2005). Fluorescent nanocrystal-polymer composites from aqueous nanocrystals: methods without ligand exchange, *Chem. Mater.*, **17**, pp. 4783–4788.

72. Sun, H., Zhang, J., Zhang, H., Li, W., Wang, C., Li, M., Tian, Y., Zhang, D., Chen, H., and Yang, B. (2006). Preparation of carbazole-containing amphiphilic copolymers: an efficient method for the incorporation of functional nanocrystals, *Macromol. Mater. Eng.*, **291**, pp. 929–936.

73. Yang, Y., Tu, C., and Gao, M. (2007). A general approach for encapsulating aqueous colloidal particles into polymeric microbeads, *J. Mater. Chem.*, **17**, pp. 2930–2935.

74. Cao, X., Li, C. M., Bao, H., Bao, Q., and Dong, H. (2007). Fabrication of strongly fluorescent quantum dot-polymer composite in aqueous solution, *Chem. Mater.*, **19**, pp. 3773–3779.

75. Luo, X., Han, J., Ning, Y., Lin, Z., Zhang, H., and Yang, B. (2011). Polyurethane-based bulk nanocomposites from 1-thioglycerol-stabilized CdTe quantum dots with enhanced luminescence, *J. Mater. Chem.*, **21**, pp. 6569–6575.

76. Guerrero-Martínez, A., Pérez-Juste, J., and Liz-Marzán, L. M. (2010). Recent progress on silica coating of nanoparticles and related nanomaterials, *Adv. Mater.*, **22**, pp. 1182–1195.

77. Jing, L., Yang, C., Qiao, R., Niu, M., Du, M., Wang, D., and Gao, M. (2010). Highly fluorescent CdTe@$SiO_2$ particles prepared *via* reverse microemulsion method, *Chem. Mater.*, **22**, pp. 420–427.

78. Yang, Y., and Gao, M. (2005). Preparation of fluorescent $SiO_2$ particles with single CdTe nanocrystal cores by the reverse microemulsion method, *Adv. Mater.*, **17**, pp. 2354–2357.

79. Yang, Y., Jing, L., Yu, X., Yan, D., and Gao, M. (2007). Coating aqueous quantum dots with silica *via* reverse microemulsion method: toward size-controllable and robust fluorescent nanoparticles, *Chem. Mater.*, **19**, pp. 4123–4128.

80. Zhang, Y., Li, Y., and Yan, X. P. (2009). Aqueous layer-by-layer epitaxy of type-II CdTe/CdSe quantum dots with near-infrared fluorescence for bioimaging applications, *Small*, **5**, pp. 185–189.

81. Zeng, Q., Kong, X., Sun, Y., Zhang, Y., Tu, L., Zhao, J., and Zhang, H. (2008). Synthesis and optical properties of type II CdTe/CdS core/shell quantum dots in aqueous solution *via* successive ion layer adsorption and reaction, *J. Phys. Chem. C*, **112**, pp. 8587–8593.

82. Lide, D. R., ed. (2003–2004). *CRC Handbook of Chemistry and Physics*, 84th Ed. (CRC Press).

83. Harrison, M. T., Kershaw, S. V., Burt, M. G., Eychmüller, A., Weller, H., and Rogach, A. L. (2000). Wet chemical synthesis and spectroscopic study of CdHgTe nanocrystals with strong near-infrared luminescence, *Mater. Sci. Eng. B*, **69**, pp. 355–360.

84. Rogach, A. L., Eychmüller, A., Hickey, S. G., and Kershaw, S. V. (2007). Infrared-emitting colloidal nanocrystals: synthesis, assembly, spectroscopy, and applications, *Small*, **3**, pp. 536–557.

85. Semonin, O. E., Johnson, J. C., Luther, J. M., Midgett, A. G., Nozik, A. J., and Beard, M. C. (2010). Absolute photoluminescence quantum yields of IR-26 dye, PbS, and PbSe quantum dots, *J. Phys. Chem. Lett.*, **1**, pp. 2445–2450.

86. Liang, G.-X., Gu, M.-M., Zhang, J.-R., and Zhu, J.-J. (2009). Preparation and bioapplication of high-quality, water-soluble, biocompatible, and near-infrared-emitting CdSeTe alloyed quantum dots, *Nanotechnology*, **20**, pp. 415103.

87. Liang, G.-X., Li, L.-L., Liu, H.-Y., Zhang, J.-R., Burda, C., and Zhu, J.-J. (2010). Fabrication of near-infrared-emitting CdSeTe/ZnS core/shell quantum dots and their electrogenerated chemiluminescence, *Chem. Commun.*, **46**, pp. 2974–2976.

88. Wuister, S. F., de Mello Donega, C., and Meijerink, A. (2004). Influence of thiol capping on the exciton luminescence and decay kinetics of CdTe and CdSe quantum dots, *J. Phys. Chem. B*, **108**, pp. 17393–17397.

89. Kotov, N. A., and Stellacci, F. (2008). Frontiers in nanoparticle research: toward greater complexity of structure and function of nanomaterials, *Adv. Mater.*, **20**, pp. 4221–4222.

# Chapter 3

# Assemblies of Thiol-Capped CdTe Nanocrystals

**Nikolai Gaponik**

*Physical Chemistry, TU Dresden, Bergstr. 66b, 01062, Dresden, Germany*

nikolai.gaponik@chemie.tu-dresden.de

Nanoparticles belong to the class of materials which is a focus of modern nanochemistry [1]. Nanoparticles may consist of organic (i.e., polymer) and inorganic (i.e., semiconductor, metal or metal oxide) materials. The development of nanochemistry in the direction of multifunctional materials gives rise to more and more sophisticated nanoparticles consisting of inorganic–inorganic, polymer–inorganic, etc., composites of various structures: core shell (also multiple shells), dumbbells, doped materials and alloys. One of the main nanochemistry approaches that allow one to obtain nanoparticles is their colloidal synthesis [2–5]. The main feature of the colloidal synthesis is the formation of nanoparticles by strictly controlled chemical precipitation in the presence of surfactants (also named ligands, capping agents, stabilizers). As it follows from the definition of a colloid, successful synthesis

*Cadmium Telluride Quantum Dots: Advances and Applications*
Edited by John F. Donegan and Yury P. Rakovich

ISBN 978-981-4316-05-7 (Hardcover), 978-981-4316-09-5 (eBook)
www.panstanford.com

should result in the formation of stable colloidal solution of nanoparticles. Besides the synthesis, post-preparative treatments of the nanoparticles (and their solutions) are very important chemical steps allowing transformation of "bare" nanoparticles into powerful, functional nano-building blocks demanded by emerging fields of nanophotonics and nanotechnology. Typical post-preparative treatments approaches include size-selective precipitation, chemical and photochemical etching, surfactant and solvent exchange, as well as mixing and forming hybrids or bulk composites with other molecules and substances.

However, building blocks themselves, even the most desired and promising from the point of view of their magic functionalities, are useless if one could not properly handle them. Indeed, to fabricate nanoparticle in solution and to functionalize it are only half of the job. To make the existing particular nanoparticle useful, we have to localize it, put it in desired architecture (often with nanometre resolution!) and finally be able to address it. At present, the most successful examples of proper handling are based on so-named top-down nanotechnology approaches: the nanoparticles themselves, their superstructures and the nano-device architectures are fabricated by e-beam epitaxy, nanolithography, etc. The methods are precise but have their limitations: They are relatively expensive, yield relatively small amount of nano-objects and also applicable to a limited list of materials and substrates. Without negating the importance of top-down approaches, it is obvious that nanotechnology may only win from the development of alternative (e.g., bottom up) fabrication and assembly approaches. Indeed, the above-mentioned limitations seem not to be inherent to typical products of bottom-up synthesis, namely to colloidally synthesized nanoparticles. Indeed, 100 mL of typical colloidal solution of the type relevant to the present work contains an average of up to $10^{17}$ of the functional fluorescent semiconductor nanoparticles. The cost of such a synthesis is below €100. Consequently, the research fields devoted to efficient synthesis and handling of the colloidal nanoparticles have quickly emerged in the last few decades. The main tools allowing the manipulation of colloidal nanoparticles are assembly and self-assembly techniques.

Nowadays assembly approaches are recognized as being the main working tool of bottom-up chemical nanotechnology. Colloidal semiconductor and metal nanocrystals (NCs) are used to build up artificial molecules and solids [6–8]. The assembly of nanocrystals can be performed on surfaces of various geometries, i.e., flat, porous and spherical. Such assemblies may be very useful for thin-film technologies, doping of mesoporous materials, modification of pre-patterned substrates, creation of microshells and fabrication of microcavities [9]. Self-assembly approaches or the use of removable templates make possible the formation of nanowires [10,11], nanosheets [12] or nanoporous 3D ordered materials [13] created solely from the assembled nanoparticles. Hierarchical assembling and assembling of nanocrystals with other functional (organic or inorganic) entities opens up the possibility to achieve composites with literally unlimited functionalities. The understanding and governing of charge and energy transfer processes between the components of the composites are the key points in their efficient utilization as building blocks in novel types of LEDs, photovoltaic and photonic devices and various optical sensors [9]. In this chapter, the recent advances made in the assembling of the colloidal thiol-capped CdTe NCs for possible applications in nanophotonics and optoelectronics are discussed.

## 3.1 Building Blocks and Thiol-Capping

As has been already described in Chapter 2, light-emitting colloidal CdTe can successfully be synthesized in aqueous solution in the presence of short-chained thiols as stabilizing agents [14–18]. Due to quantum confinement, the photoluminescence (PL) and absorption of the NCs are dependent on their size, which in most cases can be controlled by the preparative conditions [19]. The PL of the thiol-capped NCs covers, depending on their size a very broad spectral region from green (ca. 500 nm) to the near infrared (ca. 800 nm) [17]. After synthesis, CdTe NCs can be treated to improve their size distribution, PL quantum efficiency (QE), processability and stability [16]. The PL QEs of state-of-the-art thiol-capped CdTe NCs reach 60–80% [17,20,21]. The solubility of thiol-capped CdTe NCs is not limited to aqueous solutions. NCs synthesized in water can be transferred into non-polar organic solvents, such as toluene, styrene and chloroform *via* a stabilizer exchange to long-chain thiols [22], by utilizing polymerizable surfactants [23] or by employing amphiphilic molecules for the stabilization [24,25]. Being transferred to organic solutions, the NCs may be used for the fabrication of functional and processable polymer–NC composites [22,23] suitable for applications, e.g., in LEDs [26].

Short-chained thiols determine the chemical functionality and surface charge of the NC surface. This opens the possibility to manipulate these NCs, e.g., by the layer-by-layer (LbL) technique, oriented attachment, electrostatic assembly or by electrophoresis. Specific chemical functionality (e.g., amino- or carboxylic groups) is attractive for the covalent linking to the macrosurfaces (e.g., electrodes or substrates for optical studies), to the surfaces of other NCs as well as to (bio)molecules. By this, proper thiol-capping may be considered a key factor in determining the successful fabrication and application of various assemblies discussed below [9,27].

## 3.2 Layer-by-Layer Assembly

Uniform and multicomponent thin films consisting of functional molecules and/or NCs may be formed by applying the so-called LbL technique, which was originally introduced for the assembly of polymer electrolytes [28] and small particles [29] and which was recently adapted for the deposition of charged NCs both on flat [30,31] and on curved surfaces [32,33]. The formation of monolayers of deposited materials is based on the electrostatic interaction between the NCs and the surface (Fig. 3.1). Alternation of the sign of the charges of the species to be deposited allows the growth of quite thick (up to hundreds of nanometres) multilayers, while the introduction of new components in one of the layers yields the opportunity of virtually unlimited but controllable variations of LbL-multistructure compositions [34,35]. The application of the method described to the modification of artificial opals allows one to obtain intrinsically light-emitting photonic crystals which were successfully used for the investigation of photonic confinement phenomena in opals and heteroopals [36–38]. The thickness of the LbL film depends linearly on the number of deposited NC/polyelectrolyte "bilayers". Moreover, a linear dependence of the absorption in the region of the first absorption maximum on the number of bilayers was observed [39].

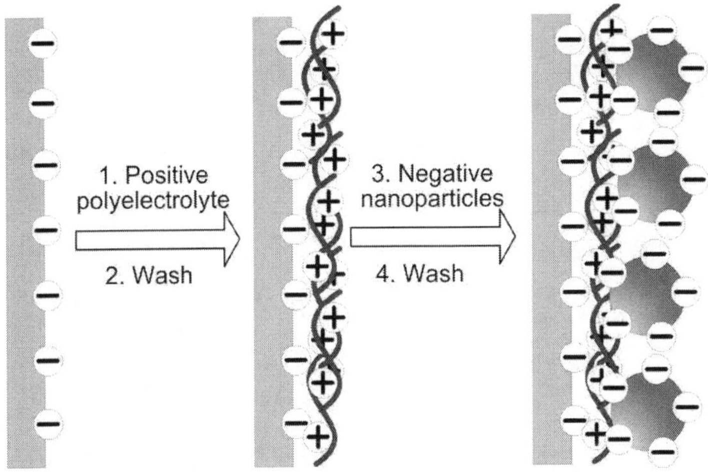


**Figure 3.1** Schematic representation of the LbL assembly involving polyelectrolyte molecules and oppositely charged nanoparticles. The procedures 1–4 can be repeated to assemble more PE/nanoparticle bilayers. Adapted with permission from ref. [27]. Copyright 2005, WILEY VCH Verlag GmbH & Co. KGaA.

Probably, the most impressive example of the application of the LbL technique is the fabrication of LEDs as was successfully demonstrated by several groups [30,40,41] and recently reviewed in ref. [42]. LEDs built by the LbL technique are normally assembled using simple wet chemical processes at room temperature in ambient conditions. The typical device, built of CdTe NCs alternatively assembled with poly(diallyldimethylamm onium chloride) (PDDA), showed an electroluminescence turn on at 2.5 V and the maximum light output at 3.3 V, with a peak radiated power of 141 nW corresponding to an external quantum efficiency of 0.51%. The luminous efficiencies of the red-emitting LED reached 0.4 $CdA^{-1}$ and 0.81 $LmW^{-1}$ [40]. In the case of patterned conducting substrates, the use of an electric field enables the acceleration or, if necessary, inhibition of the LbL assembly, which was successfully utilized for the selective fabrication of green and red emitting micro-LEDs with 50 μm spatial resolution [43]. Further development of the LEDs and LED architectures resulted in layer-by-layer fabrication of solely inorganic devices. Negatively charged CdTe NCs were intercalated within a laminar hydrotalcite-like structure instead of polyelectrolyte matrix. The laminar scaffold acted to both support and distribute the CdTe nanocrystals. The device had high thermal stability, operating continuously up to 90°C, and a maximum efficiency at 0.12 A $cm^2$ with the peak luminescence efficiency of 8.42 Cd $A^{-1}$. This thermostable device is targeted at the automotive industry [44]. Operating dashboard segments fabricated by this technology are shown in Fig. 3.2.

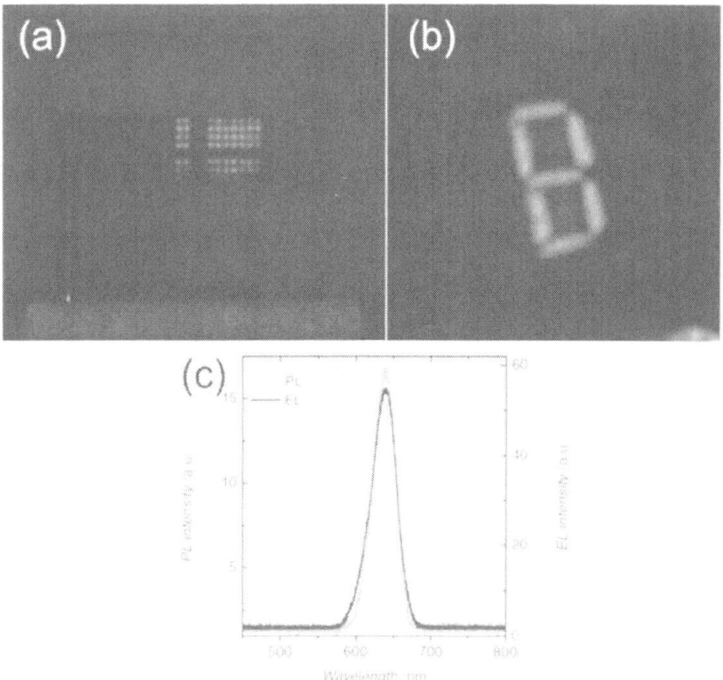


**Figure 3.2** (a) Array of pixels fabricated by the all-inorganic layer-by-layer assembly. Each pixel has dimensions of 250 μm × 250 μm; (b) "seven-segments" device fabricated with the same method, the dimension of the luminescent area is 2 mm × 7 mm; (c) photoluminescence spectrum of aqueous CdTe QDs (gray curve) compared to the electroluminescence spectrum of the device (black curve). Adapted with permission from ref. [44]. Copyright 2010, WILEY VCH Verlag GmbH & Co. KGaA.

The building of LbL films from NCs of two different sizes allows controllable design of Förster resonance energy transfer (FRET) based devices. The FRET rates as high as 4 $ns^{-1}$ were observed between CdTe NCs separated by one polyelectrolyte monolayer [45]. This phenomenon was utilized for the optimization of energy scavenging systems [46] as well as for the fine tuning of the chromacity coordinates of NC composites [47]. Varying the distance between layers of differently sized NCs influences the relative intensity of corresponding emission bands and results in controllable tuning of chromacity coordinates. The use of the LbL technique for the formation of CdTe NC and gold nanoparticle layers with controllable interlayer distance allowed an observation of various phenomena related to efficient interactions between emitting nanocrystals and surface plasmon field. Among others are off-resonance enhancement of bandgap emission [48], anisotropic emission from multilayered plasmon resonator nanocomposites of isotropic CdTe NCs [49], and surface plasmon-enhanced FRET between donor and acceptor CdTe NCs [50].

## 3.3 Electrostatic Assembly of Thin Films and Solids

This method is based on the fact that the proper choice of stabilizing agent allows one to vary the surface charge of NCs. By this the LbL-like assembly may be implemented also without utilization of polyelectrolytes or other charged substances. Successful thin layer fabrication was demonstrated in the example of cysteamine (CA) (positively charged) and TGA (negatively charged) stabilized CdTe NCs. However, it has been shown that the formation of only one bi-layer of oppositely charged NCs is achievable without using polyelectrolytes [51]. Nevertheless, even in these bilayer structures the strong electrostatic interaction between the NCs allows the observation of an increase in the FRET rates of up to 14 $ns^{-1}$. Thin films of colloidal CdTe NCs built-up by this technique were also successfully utilized for the fabrication of photodetectors [52].

Electrostatic interactions as well as covalent linking between positively charged CA stabilized NCs and negatively charged TGA stabilized CdTe NCs were utilized for the formation of closely packed solids [53,54]. The CdTe(CA)–CdTe(TGA) coupling was dictated by the surfactant spacer, ranging between 0.93 and 1.14 nm and by electrostatic and covalent interactions, enabling a FRET process among the NCs. It was demonstrated that covalent bonding allows a closer surface-to-surface distance and by this a quicker energy transport (in a range of 1 $ns^{-1}$) from smaller (donors) to larger (acceptors) NCs in assemblies [54].

Divalent positively charged $Ca^{2+}$ ions can act as electrostatic chelating linkers able to induce clustering of CdTe NCs through coordination to the negatively charged carboxylic groups of TGA or MPA ligands [55]. Moreover, the addition of sodium carbonate, which binds $Ca^{2+}$, reverses the clustering process. An efficient FRET was found to be responsible for changes in the emission spectra and the luminescence decay times of binary- and ternary-sized CdTe NC clusters bound by $Ca^{2+}$ ions.

## 3.4 Thin Film Assemblies on Microbeads

The LbL assembly of NCs on the surface of polymer (polystyrene (PS), polymethylmetacrylate (PMMA), melamine formaldehyde (MF), etc.) or silica beads opens up an opportunity to create submicronsized objects exhibiting the properties of their nano-sized components [56,57]. In this case, the LbL assembly can be done by suspension of the beads in solutions of the corresponding layer component (Fig. 3.3). The compositions of the solutions are generally the same as for the procedures described above. Repetitive centrifugation from pure water suspensions is used for washing of the beads and removal of excessive or non-specifically adsorbed reactants. The modified beads retain the PL properties of the NCs used with a slight red shift of the PL maximum due to energy transfer from the smaller to the largest NCs in the assembled surface film [33,58]. The comparatively large size of the beads allows their manipulation by optical tweezers or their use when mounted at tip on the end of an optical fibre. Thus, the beads can be used as sub-wavelength-sized light sources for photonic applications [59] or for the FRET-based selective amplification of a signal in scanning near-field optical microscopy (SNOM) [60].

The LbL modified beads of sizes above 2 μm show an efficient coupling of the light emitted by the NCs with the whispering gallery modes (WGM) of spherical microresonators [61–63]. Moreover, optical coupling between two adjacent microresonators modified with light-emitting NCs allowed observation of photonic molecule phenomena, i.e., transition from individual (atom-like) to collective (molecule-like) photonic states [62,64].

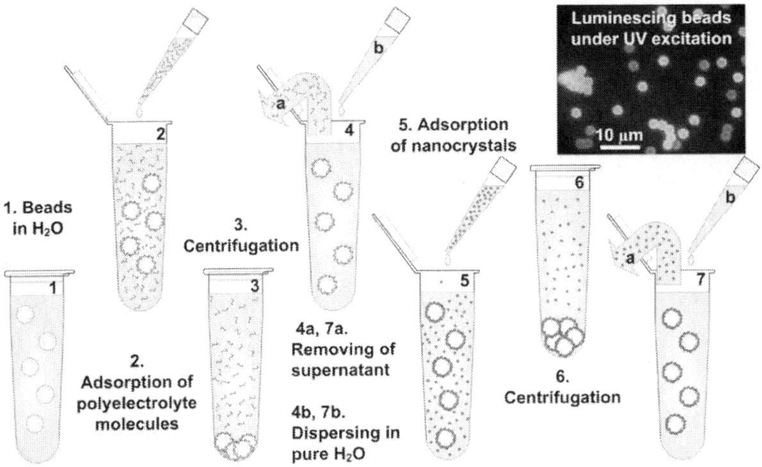


**Figure 3.3** Schematics of the layer-by-layer modification of micro and submicro beads with nanocrystals. Inset shows confocal microscopy image of the luminescing beads of three different colours.

The use of beads made from materials which can be dissolved after LbL deposition of the polyelectrolytes (such as e.g., MF, $MnCO_3$, etc.) allows the formation of hollow polyelectrolyte/NCs shells [65–67]. As shown in Fig. 3.4, by proper choice of assembled NCs the shells may be coded by different emission colours and colour combinations, or may be simultaneously magnetic (due to assembled $Fe_3O_4$ NCs) and light emitting (due to CdTe NCs) [66]. The shells are promising as microcapsules for drug delivery systems where the PL of inserted NCs can be used for the monitoring of the delivery process [68,69].

Another method of coating microspheres has been established through the controlled precipitation of NCs by means of their aggregation in solution onto latex spheres [70]. This approach is a variation of heteroaggregation phenomenon in a colloidal environment. The aggregation of NCs onto colloidal beads was induced by the solvent-non-solvent pair precipitation technique which is normally successfully used to separate semiconductor nanoparticles by size. Relatively large (from several tens of nanometres to several tens of micrometres) colloidal latex spheres statistically harvest aggregated NCs, which consequently leads to the formation of a shell of NCs on a latex core. Moreover, the thickness of the shell can efficiently be controlled through the proper choice of the concentration ratio of the colloidal spheres serving as the cores and the precipitated NC species. This method in comparison to the LbL technique has a great advantage as being a quick one-step procedure for the formation of relatively thick shells.

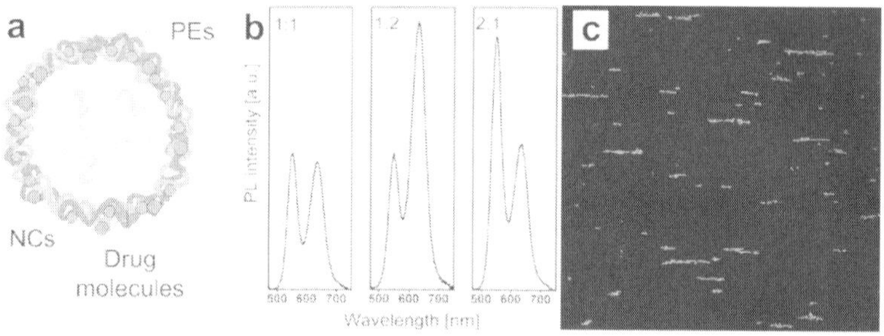


**Figure 3.4** Schematics of a hollow polymer capsule formed from polyelectrolyte (PE) bilayers. The capsule is loaded with drug molecules and labelled with two types (luminescent–luminescent or luminescent–magnetic) of nanocrystals. (b) PL spectra of two-colour coded microcapsules showing two separate peaks with controlled intensity ratios (1:1, 1:2 and 2:1); (c) Confocal microscopy image of aligned in magnetic field shells (microcapsules) (diam. 5.6 μm) simultaneously loaded with CdTe and $Fe_3O_4$ NCs. Panel (b) is adapted with permission from ref. [67]. Copyright 2002 WILEY VCH Verlag GmbH & Co. Panel (c) is adapted from ref. [66] Copyright 2004, with permission from the American Chemical Society.

## 3.5 Covalent Coupling of NCs

While the covalent coupling between thiol-capped NCs aiming to create molecular like NC species (e.g., NC dimers, etc.) is still in its infancy, carbodiimide chemistry may well be applied to the coupling of NCs to pre-treated macrosurfaces. Accordingly, CdTe NCs have been used for the conjugation with various substrates such as glass, silica or silicon with different shapes of the surfaces [71]. The method is based on the binding of acid-stabilized NCs to aminated surfaces with the aid of carbodiimide as a mediator and is not limited to bulk substrates but is of more generality as it was successfully applied to coat glass spheres of micron size or silica beads by CdTe NCs [71]. The luminescence from the modified spheres kept in air and under daylight was not altered over a period of at least 3 months.

Recently, an approach for immobilizing CdTe NCs on thin, fluid films consisting of poly(acrylic acid) (PAA) polymer chains tethered by one end to an underlying substrate in a polymer brush configuration was demonstrated [72]. The protocol involves covalent bonding similar to that described above, but between amino groups present on the surface of the CdTe NCs (stabilized by CA) and carboxylic groups of PAA polymer chains. The thickness of the PAA polymer brush employed (ca. 4.5 nm) was comparable to the nanoparticle size. Exploitation of such a thin polymer brush was found to suppress nanoparticle aggregation and facilitate complete surface coverage [72,73]. As a fine example of the practical application of covalently bound assemblies, the thermosensitive poly(*N*-isopropyl acrylamide) (PNIPAM) microgel linked to CA-capped CdTe NCs should be mentioned. Due to the temperature-dependent transition between collapsed and swollen stages of the microgel (ca. 35°C), the composite shows pronounced and reversible quenching/antiquenching fluorescence behaviour (Fig. 3.5) [74]. Such microsensors may be utilized for the monitoring of temperature changes in biological systems or in microfluidics.

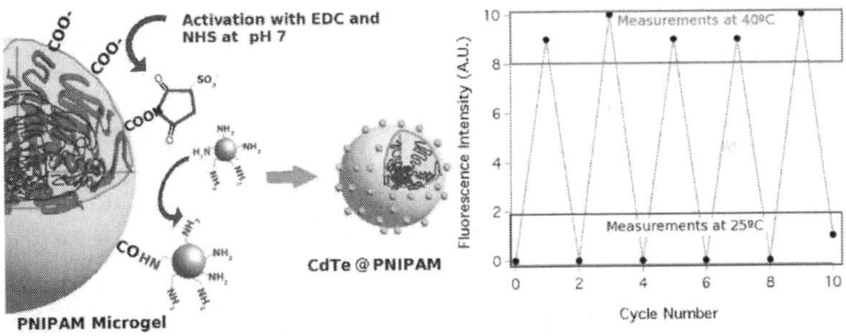


**Figure 3.5** Left: Schematics of the reaction steps carried out to obtain microgels covered with CdTe NCs. Right: maximum PL intensity measured during repeated heating (40°C) and cooling (25°C) cycles of the microgels. Adapted with permission from ref. [74]. Copyright 2008, American Chemical Society.

## 3.6 Self-Assembly on Gold

A self-assembly of CdTe NCs capped with TGA or MPA on a gold surface was recently demonstrated and was successfully utilized for electrochemical investigations [75]. For an efficient assembly proper cleaning and activation of the gold surface must be done. For this, the surface of the gold plates were polished by diamond paste followed by a treatment in boiling concentrated $HNO_3$ and $H_2SO_4$. Then, the gold plates were thoroughly washed with doubly distilled water and annealed at 700°C in air. The immersion of these plates in an aqueous solution of NCs for approximately 5 min and subsequent washing in pure water resulted in the formation of a self-assembled film of the NCs with approximately 50% surface coverage (Fig. 3.6). Apart from the mentioned electrochemical studies, this technique may be applied for the modification of gold micro- and nanostructures to be used in nanoplasmonics.

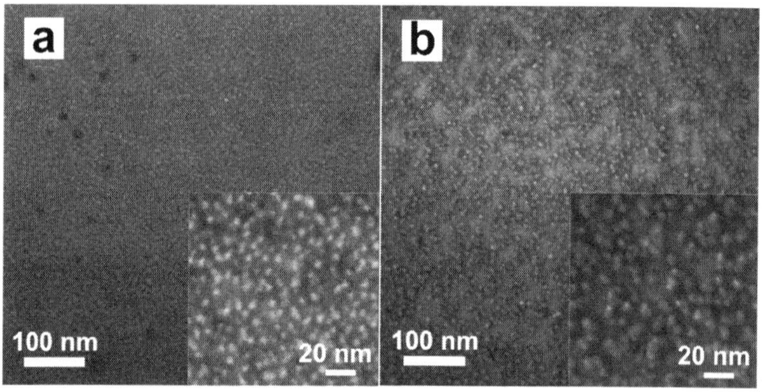


**Figure 3.6** High-resolution SEM images of the surface of a Au electrode after adsorption of CdTe nanocrystals stabilized with TGA (a) and MPA (b). The mean particle sizes were 3.3 and 7 nm, respectively. Adapted with permission from ref. [75]. Copyright 2005, American Chemical Society.

## 3.7 1D Assemblies and 2D Nanosheets

As has been shown recently, the gentle removal of TGA surfactant [76] or the partial destabilization of a colloidal solution by the addition of phosphate buffer at physiological concentrations [77,78] causes a spontaneous assembly of CdTe NCs into brightly emitting nanowires. Moreover, the latter method allowed the *in situ* observation of the process of nanowire formation under a confocal microscope. Due to this *in situ* observation, the proper conditions for a quick (in minutes) and controllable assembly were established [78]. The photoluminescence maximum of the

CdTe nanowires is commonly red-shifted by approximately 10 nm, which points towards a decrease of the electronic confinement in these 1D objects [78]. The wires were also found as a by-product of ageing CdTe NC solutions by keeping the NCs for more than 1 year at room temperature. This wire formation can be a result of slow degradation of the stabilizing thiols. Alternatively, the strongly emitting branched 1D assemblies of CdTe NCs were also prepared by a soft templating method with amphiphilic molecules [79]. Self-assembled CdTe nanowires were used for further decoration with $SiO_2$, resulting in complex nanostructures ("nanocentipedes") [80]. When similar "nanocentipedes" were formed from CdTe NCs bioconjugated *via* streptavidin-biotin coupling to CdTe NWs and conditions of optimal distances and spectral matching were achieved, an efficient cascade energy transfer resembling the principles of biological photosynthetic systems was observed (Fig. 3.7) [81]. A prototype of logical nanocircuitry formed from the triangular arrangement of bioconjugated nanowires was also demonstrated [82]. Not only wires but also strongly emitting nanotubes may be formed if CdTe NCs are assembled on the surface of Cd-thiolate nanowires [83].

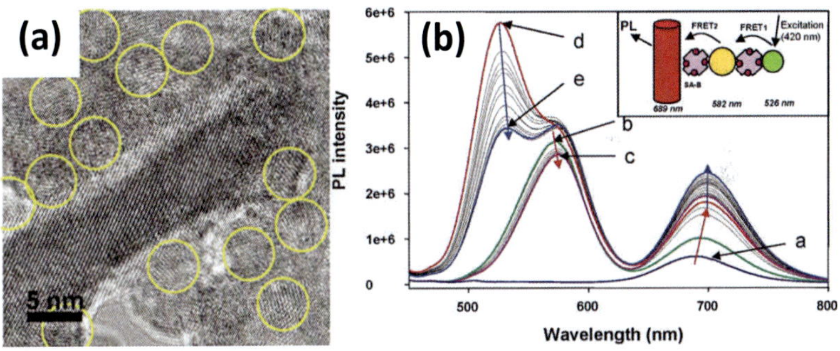


**Figure 3.7** (a) High resolution TEM image of a CdTe nanowire conjugated with CdTe nanocrystals (yellow circles indicate CdTe NCs); (b) PL spectra of multiconjugated NW-NC superstructures: a, NW only; b, when orange-emitting NCs were added; c, after 21 min; d, when green-emitting NCs were added; e, after 30 min. Arrows indicate the progression of spectra with time as bioconjugation reactions proceed. Red arrows correspond to the initial formation of NW-orange NCs conjugates. Blue arrows correspond to spectral changes started after subsequent addition of green-NCs. Adapted with permission from ref. [81]. Copyright 2005, American Chemical Society.

An interesting example of NCs assembly is the spontaneous formation of free-floating sheets which occurs when 2-(dimethy lamino)ethanethiol (DMAET) is used as the capping agent [12]. Comparative analysis of the self-assembly behaviour of TGA- and DMAET-capped CdTe NCs, which form nanowires and nanosheets, respectively, showed that NC shape, directional attraction, and electrostatic interactions are key to determining the anisotropy of the NC–NC interaction and the final self-assembled structures. Surprisingly, these data show the resemblance of NC assemblies to self-ordering biological systems, such as S-layer–forming proteins [84].

## 3.8 3D Assembled Gels and Aerogels

Although the fabrication of aerogels, highly porous materials bridging in their properties the macro- and nano dimensions [85], was known for metal oxides since 1931 [86], first demonstrations of aerogels built solely from colloidal semiconductor NCs appeared only recently [13,87,88]. In the case of thiol-capped NCs, it was shown that controllable ageing, chemical and photochemical treatment may lead to the formation of 3D networks (gelation). Photochemical treatment is shown to be the most controllable and reproducible way for gel formation, allowing to preserve the pristine emission of the NCs also in the gels [88,89]. The formation of highly porous (pore surface of ca. 210 $m^2/g$ and monolith density of less than 1/500th of corresponding bulk material) strongly emitting aerogels as well as the hybridization of the wet gels with polymers has been demonstrated (Fig. 3.8) [88]. Their hybrids with polymers are processable by common casting or thermoplating techniques. Utilization of conducting polymers for the gel infiltration will open new possibilities for fabrication, depending on design, of highly efficient hybrid LEDs or energy harvesting systems. In such composites percolated polymer and nanocrystalline phases may allow efficient separated charge carrier transport, while a huge interface area may ideally serve for carrier separation or recombination depending on design. The high porosity of the light-emitting aerogels may be utilized for optical sensing and photocatalysis.

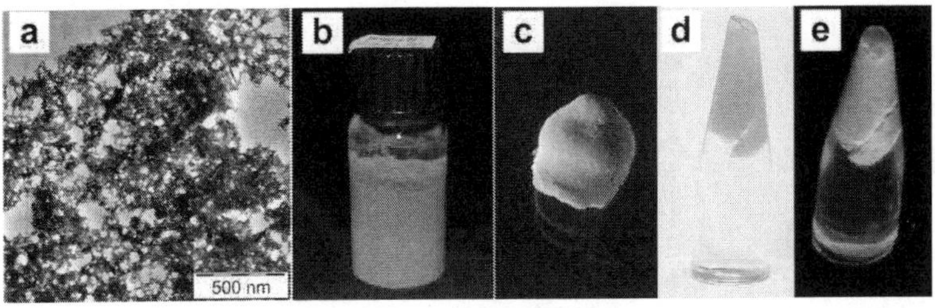


**Figure 3.8** (a) TEM image of a gel structure; (b) strongly emitting CdTe NC hydrogel; (c) self-supporting emitting CdTe aerogel (size is ca. 0.5 $cm^3$); (d, e) optically transparent and processable aerogel-PMMA composite. Adapted with permission from ref. [88]. Copyright 2008, WILEY VCH Verlag GmbH & Co. KGaA.

The method of self-assembly for CdTe nanocrystal 3D networks supported by Cd-thiolates (Fig. 3.9) was introduced recently [90]. In the resulting gel-like structures and aerogels the optical properties of the individual nanoparticles are better prevented from the deterioration: the nanoparticles are separated by the thiolate network and quenching of the emission due to the energy transfer between nanocrystals is less probable. The aerogels of this type show very strong emission and possess "saturated" pure emission colours. They may find applications in colour conversion layers [91].

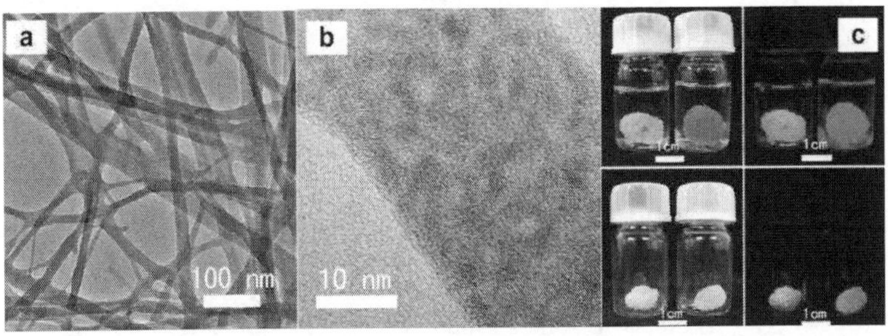


**Figure 3.9** Typical TEM (a) and HRTEM (b) images of CdTe nanocrystals–Cd thiolate complex hybrid nanostructures; (c) true colour images of the nanostructures in form of gels (top) and aerogels (bottom) under UV (right) and day light (left) illumination. Adapted with permission from ref. [90]. Copyright 2010, American Chemical Society.

Recently developed stabilization of CdTe NCs with tetrazole derivatives allowed to propose an alternative method for the formation of gel networks based on complex bridges with divalent metal ions, e.g., $Cd^{2+}$ or $Ni^{2+}$ (Fig. 3.10) [92]. This gelation is very quick (on the time scale of several seconds to several minutes) and also shown to be reversible if stronger complexants such as EDTA are added to the network. This reversibility may be easily followed by visual observations and also spectroscopically as the gel is characterized by pronounced scattering in the UV-Vis absorbance and by a red-shifted photoluminescence with relatively lower intensity. The method was also successfully utilized for the formation of mixed CdTe and Au nanoparticle aerogels [93]. Efficient controlling of the ratio between plasmonic and light-emitting nanoparticles in such kind of aerogels allowed conditions for surface plasmon enhancement of the emission of the NCs to be found [93].

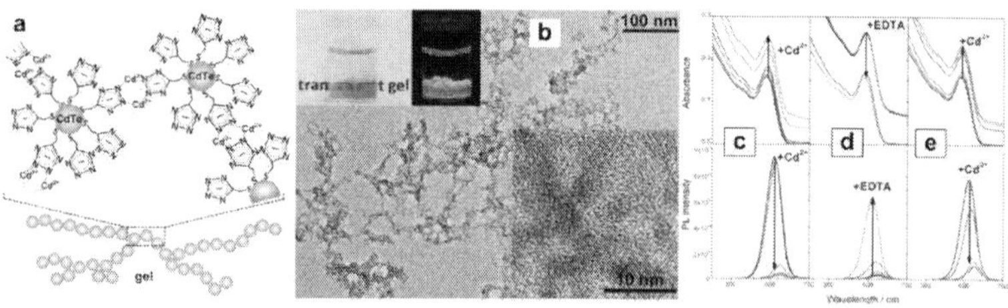


**Figure 3.10** (a) Mechanism proposed for the gelation of CdTe/5-$HSCH_2Tz$ nanocrystals; (b) TEM and HRTEM images of a CdTe nanocrystal hydrogel fragment dried on a TEM grid. Inset: a typical true-colour image of a CdTe hydrogel under day light (left) and under UV lamp (right); (c–e) absorbance (top panels) and emission (bottom panels) spectra of CdTe-Tetrazole nanocrystal colloid measured upon a stepwise addition of $Cd^{2+}$ ions into the initial colloid (c) and into the by EDTA regenerated colloid (e), (d) nanocrystal solution regeneration after step c by the addition of EDTA. The arrows indicate increased $Cd^{2+}$ or EDTA concentration. Adapted with permission from ref. [92]. Copyright 2010, American Chemical Society.

## 3.9 Conclusions and Outlook

Brightly emitting thiol-capped CdTe NCs with a flexible surface chemistry determined by easy choice of capping ligands are considered perfect building blocks for assembling by various techniques based on self-organization, electrostatic interactions, covalent linking, etc. These assemblies have already secured and will secure in the future a wide field of applications, ranging from life sciences to photonics and optoelectronics. Applications in optoelectronics cover the light-emitting diodes, microarrays of multicoloured light-emitting pixels and photosensitive films. This is closely connected to photonic applications in which these assemblies play the role of subwavelength emitters, tunable light-sources coupled to optical modes of photonic crystals and heterocrystals, spherical microresonators, photonic molecules and waveguides. Biology related applications include temperature sensitive nanoassemblies, luminescent coding of multifunctional microbeads and polymer microcapsules suitable for drug delivery systems. The ability of CdTe NCs to interact with neighbouring nanoparticles or molecules gives rise to the assembly of FRET-based and nanoplasmonic devices as well as various sensors.

It is very important to stress that although CdTe nanocrystals are the most successful and prominent members of the thiol-capped family, the methods described here are versatile and are also applicable to other types of thiol-capped NCs. Taking into account the variety of materials available in thiol-capped form (ZnSe, ZnSeTe, ZnCdSe, CdS, CdSe, CdHgTe, HgTe), each possessing unique physico-chemical properties, the above described methods as well as possible new approaches have literally unlimited applicability in nanoscience and nanotechnology.

## References

1. Schmid, G. (2010). *Nanoparticles. From Theory to Applications.* 2nd edition. Weinheim, Wiley-VCH.
2. Alivisatos, A. P. (1996). Semiconductor clusters, nanocrystals, and quantum dots, *Science*, **271**, p. 933.
3. Donegá, C. d. M., Liljeroth, P., and Vanmaekelbergh, D. (2005). Physicochemical evaluation of the hot-injection method, a synthesis route for monodisperse nanocrystals, *Small*, **1**, pp. 1152–1162.
4. Gaponik, N., Hickey, S. G., Dorfs, D., Rogach, A. L., and Eychmüller, A. (2010). Progress in the light emission of colloidal semiconductor nanocrystals, *Small*, **6**, pp. 1364–1378.

5. Weller, H. (1993). Colloidal semiconductor Q-particles: chemistry in the transition region between solid state and molecules, *Angew. Chem. Int. Ed. Engl.*, **32**, pp. 41–53.
6. Collier, C. P., Vossmeyer, T., and Heath, J. R. (1998). Nanocrystal superlattices, *Ann. Rev. Phys. Chem.*, **49**, pp. 371–404.
7. Rogach, A. L., Talapin, D. V., Shevchenko, E. V., Kornowski, A., Haase, M., and Weller, H. (2002). Organization of matter on different size scales: monodisperse nanocrystals and their superstructures, *Adv. Funct. Mater.*, **12**, pp. 653–664.
8. Shevchenko, E. V., Talapin, D. V., Kotov, N. A., O'Brien, S., and Murray, C. B. (2006). Structural diversity in binary nanoparticle superlattices, *Nature*, **439**, pp. 55–59.
9. Gaponik, N. (2010). Assemblies of thiol-capped nanocrystals as building blocks for use in nanotechnology, *J. Mater. Chem.*, **20**, pp. 5174–5181.
10. Cho, K.-S., Talapin, D. V., Gaschler, W., and Murray, C. B. (2005). Designing PbSe nanowires and nanorings through oriented attachment of nanoparticles, *J. Am. Chem. Soc.*, **127**, pp. 7140–7147.
11. Tang, Z., and Kotov, N. A. (2005). One-dimensional assemblies of nanoparticles: Preparation, properties, and promise, *Adv. Mater.*, **17**, pp. 951–962.
12. Tang, Z., Zhang, Z., Wang, Y., Glotzer, S. C., and Kotov, N. A. (2006) Self-assembly of CdTe nanocrystals into free-floating sheets, *Science*, **314**, pp. 274–278.
13. Mohanan, J. L., Arachchige, I. U., and Brock, S. L. (2005). Porous semiconductor chalcogenide aerogels, *Science*, **307**, pp. 397–400.
14. Gao, M., Kirstein, S., Möhwald, H., Rogach, A. L., Kornowski, A., Eychmüller, A., and Weller, H. (1998). Strongly photoluminescent CdTe nanocrystals by proper surface modification, *J. Phys. Chem. B*, **102**, pp. 8360–8363.
15. Gaponik, N., and Rogach, A. L. (2010). Thiol-capped CdTe nanocrystals: progress and perspectives of the related research fields, *Phys. Chem. Chem. Phys.*, **12**, pp. 8685–8693.
16. Gaponik, N., Talapin, D. V., Rogach, A. L., Hoppe, K., Shevchenko, E. V., Kornowski, A., Eychmüller, A., and Weller, H. (2002). Thiol-capping of CdTe nanocrystals: an alternative to organometallic synthetic routes, *J. Phys. Chem. B*, **106**, pp. 7177–7185.

17. Rogach, A. L., Franzl, T., Klar, T. A., Feldmann, J., Gaponik, N., Lesnyak, V., Shavel, A., Eychmüller, A., Rakovich, Y. P., and Donegan, J. F. (2007). Aqueous synthesis of thiol-capped CdTe nanocrystals: state-of-the-art, *J. Phys. Chem. C*, **111**, pp. 14628–14637.

18. Rogach, A. L., Katsikas, L., Kornowski, A., Su, D., Eychmüller, A., and Weller, H. (1996). Synthesis and characterization of thiol-stabilized CdTe nanocrystals, *Ber. Bunsen-Ges. Phys. Chem.*, **100**, pp. 1772–1778.

19. Gaponik, N., and Rogach, A. (2008). *Semiconductor Nanocrystal Quantum Dots*, ed. A. L. Rogach, (Springer-Verlag, Wien) pp. 73–99.

20. Bao, H., Gong, Y., Li, Z., and Gao, M. (2004). Enhancement effect of illumination on the photoluminescence of water-soluble CdTe nanocrystals: toward highly fluorescent CdTe/CdS core-shell structure, *Chem. Mater.*, **16**, pp. 3853–3859.

21. Grabolle, M., Spieles, M., Lesnyak, V., Gaponik, N., Eychmüller, A., and Resch-Genger, U. (2009). Determination of the fluorescence quantum yield of quantum dots: suitable procedures and achievable uncertainties, *Anal. Chem.*, **81**, pp. 6285–6294.

22. Gaponik, N., Talapin, D. V., Rogach, A. L., Eychmüller, A., and Weller, H. (2002). Efficient phase transfer of luminescent thiol-capped nanocrystals: from water to nonpolar organic solvents, *Nano Lett.*, **2**, pp. 803–806.

23. Zhang, H., Wang, C., Li, M., Ji, X., Zhang, J., and Yang, B. (2005). Fluorescent nanocrystal-polymer composites from aqueous nanocrystals: methods without ligand exchange, *Chem. Mater.*, **17**, pp. 4783–4788.

24. Dubavik, A., Lesnyak, V., Thiessen, W., Gaponik, N., Wolff, T., and Eychmüller, A. (2009). Synthesis of amphiphilic CdTe nanocrystals, *J. Phys. Chem. C*, **113**, pp. 4748–4750.

25. Smith, A. M., and Nie, S. (2008). Nanocrystal synthesis in an amphibious bath: spontaneous generation of hydrophilic and hydrophobic surface coatings, *Angew. Chem. Int. Ed.*, **47**, pp. 9916–9921.

26. Koktysh, D. S., Gaponik, N., Reufer, M., Crewett, J., Scherf, U., Eychmüller, A., Lupton, J. M., Rogach, A. L., and Feldmann, J. (2004). Near-infrared electroluminescence from HgTe nanocrystals, *Chem. Phys. Chem.*, **5**, pp. 1435–1438.

27. Shavel, A., Gaponik, N., and Eychmüller, A. (2005). The assembling of semiconductor nanocrystals, *Eur. J. Inorg. Chem.*, **2005**, pp. 3613–3623.

28. Decher, G. (1997). Fuzzy nanoassemblies: toward layered polymeric multicomposites, *Science*, **277**, pp. 1232–1237.

29. Iler, R. K. (1966). Multilayers of colloidal particles, *J. Colloid Interf. Sci.*, **21**, pp. 569–594.

30. Gao, M., Lesser, C., Kirstein, S., Möhwald, H., Rogach, A. L., and Weller, H. (2000). Electroluminescence of different colors from polycation/CdTe nanocrystal self-assembled films, *J. Appl. Phys.*, **87**, pp. 2297–2302.

31. Rogach, A. L., Koktysh, D. S., Harrison, M., and Kotov, N. A. (2000). Layer-by-layer assembled films of HgTe nanocrystals with strong infrared emission, *Chem. Mater.*, **12**, pp. 1526–1528.

32. Crisp, M. T., and Kotov, N. A. (2003). Preparation of nanoparticle coatings on surfaces of complex geometry, *Nano Lett.*, **3**, pp. 173–177.

33. Susha, A. S., Caruso, F., Rogach, A. L., Sukhorukov, G. B., Kornowski, A., Möhwald, H., Giersig, M., Eychmüller, A., and Weller, H. (2000). Formation of luminescent spherical core-shell particles by the consecutive adsorption of polyelectrolyte and CdTe(S) nanocrystals on latex colloids, *Coll. Surf. A*, **163**, pp. 39–44.

34. Klar, T. A., Franzl, T., Rogach, A. L., and Feldmann, J. (2005). Super-efficient exciton funneling in layer-by-layer semiconductor nanocrystal structures, *Adv. Mater.*, **17**, pp. 769–773.

35. Mamedov, A. A., Belov, A., Giersig, M., Mamedova, N. N., and Kotov, N. A. (2001). Nanorainbows: graded semiconductor films from quantum dots, *J. Am. Chem. Soc.*, **123**, pp. 7738–7739.

36. Gaponik, N., Eychmüller, A., Rogach, A. L., Solovyev, V. G., Sotomayor Torres, C. M., and Romanov, S. G. (2004). Structure-related optical properties of luminescent hetero-opals, *J. Appl. Phys.*, **95**, pp. 1029–1035.

37. Romanov, S. G., Chigrin, D. N., Solovyev, V. G., Maka, T., Gaponik, N., Eychmüller, A., Rogach, A. L., and Sotomayor Torres, C. M. (2003). Light emission in a directional photonic bandgap, *Phys. Stat. Sol. A*, **197**, pp. 662–672.

38. Romanov, S. G., Chigrin, D. N., Sotomayor Torres, C. M., Gaponik, N., Eychmüller, A., and Rogach, A. L. (2004). Emission stimulation in a directional band gap of a CdTe-loaded opal photonic crystal, *Phys. Rev. E*, **69**, p. 046606.

39. Ostrander, J. W., Mamedov, A. A., and Kotov, N. A. (2001). Two modes of linear layer-by-layer growth of nanoparticle-polyelectrolyte multilayers and different interactions in the layer-by-layer deposition, *J. Am. Chem. Soc.*, **123**, pp. 1101–1110.

40. Bertoni, C., Gallardo, D., Dunn, S., Gaponik, N., and Eychmüller, A. (2007). Fabrication and characterization of red-emitting electroluminescent devices based on thiol-stabilized semiconductor nanocrystals, *Appl. Phys. Lett.*, **90**, p. 034107.

41. Gallardo, D. E., Bertoni, C., Dunn, S., Gaponik, N., and Eychmüller, A. (2007). Cathodic and anodic material diffusion in polymer/semiconductor-nanocrystal composite devices, *Adv. Mater.*, **19**, pp. 3364–3367.

42. Rogach, A. L., Gaponik, N., Lupton, J. M., Bertoni, C., Gallardo, D. E., Dunn, S., Li Pira, N., Paderi, M., Repetto, P., Romanov, S. G., O'Dwyer, C., Sotomayor Torres, C. M., and Eychmüller, A. (2008). Light-emitting diodes with semiconductor nanocrystals, *Angew. Chem. Int. Ed.*, **47**, pp. 6538–6549.

43. Gao, M., Sun, J., Dulkeith, E., Gaponik, N., Lemmer, U., and Feldmann, J. (2002). Lateral patterning of CdTe nanocrystal films by the electric field directed layer-by-layer assembly method, *Langmuir*, **18**, pp. 4098–4102.

44. Bendall, J. S., Paderi, M., Ghigliotti, F., Li Pira, N., Lambertini, V., Lesnyak, V., Gaponik, N., Visimberga, G., Eychmüller, A., Sotomayor Torres, C. M., Welland, M. E., Gieck, C., and Marchese, L. (2010). Layer-by-layer all-inorganic quantum-dot-based LEDs: a simple procedure with robust performance, *Adv. Funct. Mater.*, **20**, pp. 3298–3302.

45. Franzl, T., Koktysh, D. S., Klar, T. A., Rogach, A. L., Feldmann, J., and Gaponik, N. (2004). Fast energy transfer in layer-by-layer assembled CdTe nanocrystal bilayers, *Appl. Phys. Lett.*, **84**, pp. 2904–2906.

46. Franzl, T., Klar, T. A., Schietinger, S., Rogach, A. L., and Feldmann, J. (2004). Exciton recycling in graded gap nanocrystal structures, *Nano Lett.*, **4**, pp. 1599–1603.

47. Cicek, N., Nizamoglu, S., Ozel, T., Mutlugun, E., Karatay, D. U., Lesnyak, V., Otto, T., Gaponik, N., Eychmüller, A., and Demir, H. V. (2009). Structural tuning of color chromaticity through nonradiative energy transfer by interspacing CdTe nanocrystal monolayers, *Appl. Phys. Lett.*, **94**, p. 061105.

48. Komarala, V. K., Rakovich, Y. P., Bradley, A. L., Byrne, S. J., Gun'ko, Y. K., Gaponik, N., and Eychmüller, A. (2006). Off-resonance surface plasmon enhanced spontaneous emission from CdTe quantum dots, *Appl. Phys. Lett.*, **89**, p. 253118.

49. Ozel, T., Nizamoglu, S., Sefunc, M. A., Samarskaya, O., Ozel, I. O., Mutlugun, E., Lesnyak, V., Gaponik, N., Eychmüller, A., Gaponenko,

S. V., and Demir, H. V. (2011). Anisotropic emission from multilayered plasmon resonator nanocomposites of isotropic semiconductor quantum dots, *ACS Nano*, **5**, pp. 1328–1334.

50. Lunz, M., Gerard, V. A., Gun'ko, Y. K., Lesnyak, V., Gaponik, N., Susha, A. S., Rogach, A. L., and Bradley, A. L. (2011). Surface plasmon enhanced energy transfer between donor and acceptor CdTe nanocrystal quantum dot monolayers, *Nano Lett.*, **11**, pp. 3341–3345.
51. Franzl, T., Shavel, A., Rogach, A. L., Gaponik, N., Klar, T. A., Eychmüller, A., and Feldmann, J. (2005). High-rate unidirectional energy transfer in directly assembled CdTe nanocrystal bilayers, *Small*, **1**, pp. 392–395.
52. Tu, C.-C., and Lin, L. Y. (2008). High efficiency photodetectors fabricated by electrostatic layer-by-layer self-assembly of CdTe quantum dots, *Appl. Phys. Lett.*, **93**, p. 163107.
53. Hoppe, K., Geidel, E., Weller, H., and Eychmüller, A. (2002). Covalently bound CdTe nanocrystals, *Phys. Chem. Chem. Phys.*, **4**, pp. 1704–1706.
54. Osovsky, R., Shavel, A., Gaponik, N., Amirav, L., Eychmüller, A., Weller, H., and Lifshitz, E. (2005). Electrostatic and covalent interactions in CdTe nanocrystalline assemblies, *J. Phys. Chem. B*, **109**, pp. 20244–20250.
55. Mayilo, S., Hilhorst, J., Susha, A. S., Hoehl, C., Franzl, T., Klar, T. A., Rogach, A. L., and Feldmann, J. (2008). Energy transfer in solution-based clusters of CdTe nanocrystals electrostatically bound by calcium ions, *J. Phys. Chem. C*, **112**, pp. 14589–14594.
56. Caruso, F. (2001). Nanoengineering of particle surfaces, *Adv. Mater.*, **13**, pp. 11–22.
57. Caruso, F., Caruso, R. A., and Möhwald, H. (1998). Nanoengineering of inorganic and hybrid hollow spheres by colloidal templating, *Science*, **282**, pp. 1111–1114.
58. Kagan, C. R., Murray, C. B., Nirmal, M., and Bawendi, M. G. (1996). Electronic energy transfer in CdSe quantum dot solids, *Phys. Rev. Lett.*, **76**, pp. 1517–1520.
59. Olk, P., Buchler, B. C., Sandoghdar, V., Gaponik, N., Eychmüller, A., and Rogach, A. L. (2004). Subwavelength emitters in the near-infrared based on mercury telluride nanocrystals, *Appl. Phys. Lett.*, **84**, pp. 4732–4734.
60. Müller, F., Götzinger, S., Gaponik, N., Weller, H., Mlynek, J., and Benson, O. (2004). Investigation of energy transfer between CdTe

nanocrystals on polystyrene beads and dye molecules for FRET-SNOM applications, *J. Phys. Chem. B.*, **108**, pp. 14527–14534.

61. Rakovich, Y. P., Donegan, J. F., Gaponik, N., and Rogach, A. L. (2003). Raman scattering and anti-Stokes emission from a single spherical microcavity with a CdTe quantum dot monolayer, *Appl. Phys. Lett.*, **83**, pp. 2539–2541.

62. Rakovich, Y. P., Donegan, J. F., Gerlach, M., Bradley, A. L., Connolly, T. M., Boland, J. J., Gaponik, N., and Rogach, A. (2004). Fine structure of coupled optical modes in photonic molecules, *Phys. Rev. A*, **70**, p. 051801.

63. Rakovich, Y. P., Yang, L., McCabe, E. M., Donegan, J. F., Perova, T., Moore, A., Gaponik, N., and Rogach, A. (2003). Whispering gallery mode emission from a composite system of CdTe nanocrystals and a spherical microcavity, *Semicond. Sci. Technol.*, **18**, pp. 914–918.

64. Rakovich, Y. P., Gerlach, M., Bradley, A. L., Donegan, J. F., Connolly, T. M., Boland, J. J., Przyjalgowski, M. A., Ryder, A., Gaponik, N., and Rogach, A. L. (2004). Confined optical modes in small photonic molecules with semiconductor nanocrystals, *J. Appl. Phys.*, **96**, pp. 6761–6765.

65. Gaponik, N., Radtchenko, I. L., Gerstenberger, M. R., Fedutik, Y. A., Sukhorukov, G. B., and Rogach, A. L. (2003). Labeling of biocompatible polymer microcapsules with near-infrared emitting nanocrystals, *Nano Lett.*, **3**, pp. 369–372.

66. Gaponik, N., Radtchenko, I. L., Sukhorukov, G. B., and Rogach, A. L. (2004). Luminescent polymer microcapsules addressable by a magnetic field, *Langmuir*, **20**, pp. 1449–1452.

67. Gaponik, N., Radtchenko, I. L., Sukhorukov, G. B., Weller, H., and Rogach, A. L. (2002). Toward encoding combinatorial libraries: charge-driven microencapsulation of semiconductor nanocrystals luminescing in the visible and near IR, *Adv. Mater.*, **14**, pp. 879–882.

68. Sukhorukov, G. B., Rogach, A. L., Garstka, M., Springer, S., Parak, W. J., Munoz-Javier, A., Kreft, O., Skirtach, A. G., Susha, A. S., Ramaye, Y., Palankar, R., and Winterhalter, M. (2007). Multifunctionalized polymer microcapsules: novel tools for biological and pharmacological applications, *Small*, **3**, pp. 944–955.

69. Sukhorukov, G. B., Rogach, A. L., Zebli, B., Liedl, T., Skirtach, A. G., Koehler, K., Antipov, A. A., Gaponik, N., Susha, A. S., Winterhalter, M., and Parak, W. J. (2005). Nanoengineered polymer capsules: tools for

detection, controlled delivery, and site-specific manipulation, *Small*, **1**, pp. 194–200.

70. Radtchenko, I. L., Sukhorukov, G. B., Gaponik, N., Kornowski, A., Rogach, A. L., and Möhwald, H. (2001). Core-shell structures formed by the solvent-controlled precipitation of luminescent CdTe nanocrystals on latex spheres, *Adv. Mater.*, **13**, pp. 1684–1687.
71. Shavel, A., Gaponik, N., and Eychmüller, A. (2005). Covalent linking of CdTe nanocrystals to amino-functionalized surfaces, *Chem. Phys. Chem.*, **6**, pp. 449–451.
72. Gupta, S., Uhlmann, P., Agrawal, M., Lesnyak, V., Gaponik, N., Simon, F., Stamm, M., and Eychmüller, A. (2008). Covalent immobilization of quantum dots on macroscopic surfaces using poly(acrylic acid) brushes, *J. Mater. Chem.*, **18**, pp. 214–220.
73. Ionov, L., Sapra, S., Synytska, A., Rogach, A. L., Stamm, M., and Diez, S. (2006). Fast and spatially resolved environmental probing using stimuli-responsive polymer layers and fluorescent nanocrystals, *Adv. Mater.*, **18**, pp. 1453–1457.
74. Agrawal, M., Rubio-Retama, J., Zafeiropoulos, N. E., Gaponik, N., Gupta, S., Cimrova, V., Lesnyak, V., Lopez-Cabarcos, E., Tzavalas, S., Rojas-Reyna, R., Eychmüller, A., and Stamm, M. (2008). Switchable photoluminescence of CdTe nanocrystals by temperature-responsive microgels, *Langmuir*, **24**, pp. 9820–9824.
75. Poznyak, S. K., Osipovich, N. P., Shavel, A., Talapin, D. V., Gao, M., Eychmüller, A., and Gaponik, N. (2005). Size-dependent electrochemical behavior of thiol-capped CdTe nanocrystals in aqueous solution, *J. Phys. Chem. B*, **109**, pp. 1094–1100.
76. Tang, Z., Kotov, N. A., and Giersig, M. (2002). Spontaneous organization of single CdTe nanoparticles into luminescent nanowires, *Science*, **297**, pp. 237–240.
77. Rakovich, Y. P., Volkov, Y., Sapra, S., Susha, A. S., Doeblinger, M., Donegan, J. F., and Rogach, A. L. (2007). CdTe nanowire networks: fast self-assembly in solution, internal structure, and optical properties, *J. Phys. Chem. C*, **111**, pp. 18927–18931.
78. Volkov, Y., Mitchell, S., Gaponik, N., Rakovich, Y. P., Donegan, J. F., Kelleher, D., and Rogach, A. L. (2004). *In situ* observation of nanowire growth from luminescent CdTe nanocrystals in a phosphate buffer solution, *Chem. Phys. Chem.*, **5**, pp. 1600–1602.
79. Zhang, L., Gaponik, N., Müller, J., Plate, U., Weller, H., Erker, G., Fuchs, H., Rogach, A. L., and Chi, L. (2005). Branched wires of CdTe

nanocrystals using amphiphilic molecules as templates, *Small*, **1**, pp. 524–527.

80. Wang, Y., Tang, Z., Liang, X., Liz-Marzan, L. M., and Kotov, N. A. (2004). $SiO_2$-coated CdTe nanowires: bristled nano centipedes, *Nano Lett.*, **4**, pp. 225–231.

81. Lee, J., Govorov, A. O., and Kotov, N. A. (2005). Bioconjugated superstructures of CdTe nanowires and nanoparticles: multistep cascade Förster resonance energy transfer and energy channeling, *Nano Lett.*, **5**, pp. 2063–2069.

82. Wang, Y., Tang, Z., Tan, S., and Kotov, N. A. (2005). Biological assembly of nanocircuit prototypes from protein-modified CdTe nanowires, *Nano Lett.*, **5**, pp. 243–248.

83. Niu, H., and Gao, M. (2006). Diameter-tunable CdTe nanotubes templated by 1D nanowires of cadmium thiolate polymer, *Angew. Chem. Int. Ed.*, **45**, pp. 6462–6466.

84. Zhang, Z., Tang, Z., Kotov, N. A., and Glotzer, S. C. (2007). Simulations and analysis of self-assembly of CdTe nanoparticles into wires and sheets, *Nano Lett.*, **7**, pp. 1670–1675.

85. Hüsing, N., and Schubert, U. (1998). Aerogels-airy materials: chemistry, structure, and properties, *Angew. Chem. Int. Ed.*, **37**, pp. 22–45.

86. Kistler, S. S. (1931). Coherent expanded aerogels and jellies, *Nature*, **127**, p. 741.

87. Arachchige, I. U., and Brock, S. L. (2007). Sol-gel methods for the assembly of metal chalcogenide quantum dots, *Acc. Chem. Res.*, **40**, pp. 801–809.

88. Gaponik, N., Wolf, A., Marx, R., Lesnyak, V., Schilling, K., and Eychmüller, A. (2008). Three-dimensional self-assembly of thiol-capped CdTe nanocrystals: gels and aerogels as building blocks for nanotechnology, *Adv. Mater.*, **20**, pp. 4257–4262.

89. Arachchige, I. U., and Brock, S. L. (2007). Highly luminescent quantum-dot monoliths, *J. Am. Chem. Soc.*, **129**, pp. 1840–1841.

90. Chen, H., Lesnyak, V., Bigall, N. C., Gaponik, N., and Eychmüller, A. (2010). Self-assembly of TGA-capped CdTe nanocrystals into three-dimensional luminescent nanostructures, *Chem. Mater.*, **22**, pp. 2309–2314.

91. Jang, E., Jun, S., Jang, H., Lim, J., Kim, B., and Kim, Y. (2010). White-light-emitting diodes with quantum dot color converters for display backlights, *Adv. Mater.*, **22**, pp. 3076–3080.

92. Lesnyak, V., Voitekhovich, S. V., Gaponik, P. N., Gaponik, N., and Eychmüller, A. (2010). CdTe nanocrystals capped with a tetrazolyl analogue of thioglycolic acid: aqueous synthesis, characterization, and metal-assisted assembly, *ACS Nano*, **4**, pp. 4090–4096.

93. Lesnyak, V., Wolf, A., Dubavik, A., Borchardt, L., Voitekhovich, S. V., Gaponik, N., Kaskel, S., and Eychmüller, A. (2011). 3D assembly of semiconductor and metal nanocrystals: hybrid CdTe/Au structures with controlled content, *J. Am. Chem. Soc.*, **133**, pp. 13413–13420.

Chapter 4

# Förster Resonant Energy Transfer in CdTe Nanocrystal Quantum Dot Structures

**Manuela Lunz and A. Louise Bradley**

*Photonics Group School of Physics, Trinity College, Dublin, Ireland*

louise.bradley@tcd.ie

## 4.1 Introduction

Semiconductor nanocrystal quantum dots (QD) are intensively researched in terms of their synthesis, assembly and applications [1,2]. Their size tuneable, optical properties have been demonstrated in a range of material systems covering various regions of interest across the spectrum. Their surface chemistry can be well controlled and they can be assembled, using a range of methods, to form structures of any dimension, including QD molecules, clusters, chains, two-dimensional layers and solids [2,3]. Due to these unique properties and geometry possibilities, colloidal QDs are excellent building blocks for energy transfer structures and Förster resonance energy transfer (FRET) is one of the most studied mechanisms in QD assemblies [4–9]. It is observed in

*Cadmium Telluride Quantum Dots: Advances and Applications*
Edited by John F. Donegan and Yury P. Rakovich

ISBN 978-981-4316-05-7 (Hardcover), 978-981-4316-09-5 (eBook)
www.panstanford.com

structures of all dimensions assembled from monodispersed QD solutions and more complex multi-QD architectures. FRET has been seen to play a fundamental role in determining the optical properties of the QD assemblies and consequently, in the design of QD-based devices FRET must be considered as an inherent underlying process which modifies the optical properties of the collective assembly relative to the properties of the QDs in solution. Importantly, FRET can also be engineered for a broad range of biological and optoelectronic applications including sensing [10–13], light-emitting devices [14–17] and photovoltaics [18–22]. The device applications are based both on energy transfer between QDs of different sizes and on hybrid systems in which the QD acts as the energy donor or energy acceptor. A recent review of FRET in QD hybrid systems can be found in reference [23]. Aqueous synthesized thiol-capped CdTe nanocrystal QDs in particular have received a lot of interest for FRET studies and applications as they facilitate close proximity to the nanocrystal donor or acceptor. Further details of the synthesis can be found in Chapters 2 and 3 and references therein. This chapter provides a review of the fundamentals of FRET and looks at examples of FRET in CdTe QD structures scaled from two-QD molecules to QD solids. Using examples of monolayer and bi-layer structures, the suitability of the application of standard FRET theory, for describing energy transfer between nanocrystal QDs is discussed. Finally, some consideration is given to the possibility of increasing the FRET rate and interaction distance in multilayer CdTe QD structures through surface plasmon enhancement of the FRET process.

## 4.2 FRET Theory

The first theoretical description of energy transfer from a donor to an acceptor molecule by dipole–dipole interaction was published by Theodor Förster in 1948 [24]. Therefore, this type of energy transfer is often referred to as Förster resonant energy transfer (FRET). Radiative energy transfer, corresponding to the emission of a photon by the donor and its re-absorption by the acceptor, and tunnelling, due to the spatial overlap of donor and acceptor wave-functions [25,26], are the other important energy transfer mechanisms aside from FRET. Photon re-absorption can be easily distinguished from the two other energy transfer mechanisms as it does not cause a decrease of the donor photoluminescence lifetime [27,28]. Tunnelling also leads to a shortening of the donor lifetime, but distance and/or temperature dependent measurements can be carried out to differentiate between tunnelling and FRET [6,7,9,29]. Furthermore, tunnelling is a short-range energy transfer mechanism, usually only important on sub-nanometre scales [25,26], and can normally be excluded as the origin for energy transfer for donor–acceptor separations larger than 2 nm.

### 4.2.1 FRET in QD Systems

The FRET equations discussed here have been obtained from FRET theories developed for molecules that can be represented as point dipoles. Due to the large size of the QDs, the inhomogeneous broadening of the QD ensembles as well as the different nature of the (non-)radiative processes in molecules and QDs, it has not been clear for a long time if these equations could also be applied directly to QD systems [8,30]. However, recent theoretical work on FRET between QDs suggests that it is valid to use the original FRET rate equations and the dipole approximation to describe FRET, and in particular the FRET rate, for spherical, direct-gap semiconductor nanocrystals [4,8,31–33]. Furthermore, some of the experimental work on FRET in QD systems presented in this chapter shows that these FRET equations can be used to describe the optical properties of QD FRET systems [5,6,34–37]. In the following, the most important theoretical equations describing Förster resonant energy transfer—represented by the energy transfer rate ($k_{\mathrm{FRET}}$), the energy transfer efficiency ($E_{\mathrm{FRET}}$) and the characteristic distance: the Förster radius ($R_0$)—and its effect on the optical properties of the donor and acceptor species are summarized.

### 4.2.2 General Observations

As depicted schematically in Fig. 4.1a, a luminescent species "D" (the energy donor) can lose its excitation energy *via* a radiative channel with decay rate $k_r$ or non-radiative channels, giving rise to the decay rate $k_{nr}$. As shown in Eq. (4.1), these two processes determine the photoluminescence lifetime $\tau_D$ of the donor:

$$\tau_{\mathrm{D}} = \frac{1}{k_{\mathrm{r}} + k_{\mathrm{nr}}}. \tag{4.1}$$

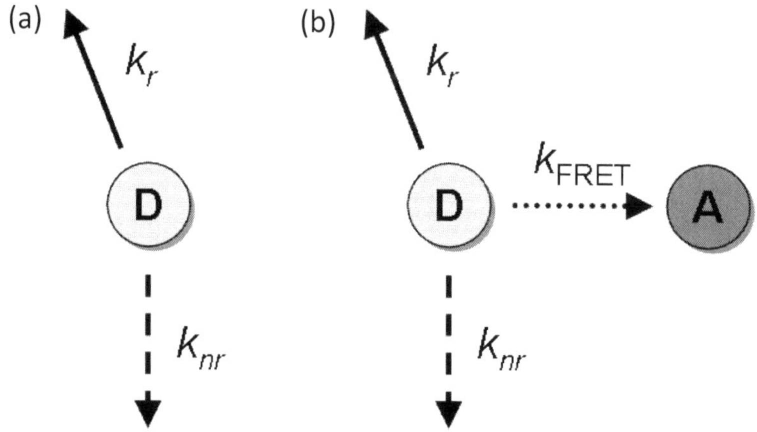


**Figure 4.1** Schematic of donor decay paths. (a) Decay paths for an excited donor (D) *via* radiative and non-radiative channels with rates $k_r$ and $k_{nr}$ respectively. (b) In the presence of a suitable acceptor (A) another decay channel *via* energy transfer, with an associated decay rate $k_{FRET}$, can lead to the de-excitation of the donor.

If a suitable acceptor "A" is present in proximity to the donor, as schematically presented in Fig. 4.1b, the excitation energy of the donor can also be transferred to the acceptor *via* energy transfer mediated by dipole–dipole interactions with a rate $k_{FRET}$. Adding another possible decay channel increases the probability for non-radiative donor de-excitation and decreases the donor emission intensity as well as its photoluminescence lifetime $\tau_{DA}$, as described in Eq. (4.2):

$$\tau_{DA} = \frac{1}{k_r + k_{nr} + k_{FRET}} = \frac{1}{\tau_D^{-1} + k_{FRET}}. \tag{4.2}$$

Energy transfer from a donor to an acceptor *via* FRET always results in a decrease of the donor lifetime and also has effects on the emission and photoluminescence lifetime of a luminescent acceptor. Typically, in a QD FRET system, both the donor and the acceptor are excited at the same time due to the broad absorption of the QDs. Energy transfer, in general, leads to an enhanced acceptor emission and the additional pumping *via* the donor, over timescales longer than the excitation pulse, and can give rise to a prolonged acceptor emission. Furthermore, energy transfer can even lead to an increase in the overall acceptor emission intensity after the excitation pulse occurred, leading to a so-called rise-time in the time-trace of the acceptor photoluminescence [9,28,38].

For broad donor and acceptor ensembles, arising mainly from the size distribution in the case for colloidal QDs, energy transfer from a sub-ensemble of donors to a particular sub-ensemble of acceptors might be more efficient than from others. Therefore, certain parts of the donor photoluminescence spectrum—corresponding to the emission from this sub-ensemble—might experience a stronger quenching than others; or a stronger enhancement in the case of the acceptors. This can lead to spectral changes in the donor and acceptor ensemble emission spectra, such as shifted peak emission wavelengths or narrowing/broadening of the emission features.

One of the prerequisites for FRET, is the presence of resonant donor and acceptor states between which the energy can be transferred. For this reason the spectral overlap $J$, as given in Eq. (4.3), is an important parameter in FRET. It describes the overlap of the donor emission $I_D(\lambda)$, using the area-normalized spectrum $\hat{I}_D(\lambda)$ in Eq. (4.3), and the acceptor absorption spectra, represented by the wavelength dependent extinction coefficient $\varepsilon_A(\lambda)$.

$$J = \int_0^\infty \hat{I}_D(\lambda) \cdot \varepsilon_A(\lambda) \cdot \lambda^4 d\lambda \tag{4.3}$$

Including $\varepsilon_A(\lambda)$ in units of $M^{-1} \cdot cm^{-1}$ and $\lambda$ in nm, $J$ is calculated in $nm^4/M \cdot cm$ with Eq. (4.3). Knowing the spectral overlap $J$ and the donor quantum yield $Q_D$, the Förster radius $R_0$, the characteristic interaction distance of FRET, can be calculated with Eq. (4.4). The result is given in nm, if $J$ is included in the calculation in $nm^4/M \cdot cm$.

$$R_0 = 0.211 \cdot \left( \frac{\kappa^2 \cdot Q_D}{n^4} \cdot J \right)^{1/6} \tag{4.4}$$

The orientation factor $\kappa^2$ is typically 2/3 for randomly oriented donor and acceptor dipoles and $n$ represents the refractive index of the surrounding medium. Typical values of $R_0$ range between 1 and 10 nm [27,28].

### 4.2.3 FRET between a Single Donor and Acceptor

Considering FRET between a single donor and a single acceptor, $R_0$ is defined as the donor–acceptor separation, $r_{DA}$, at which the FRET rate, $k_{FRET}$, is equal to the other donor decay rates $\tau_D^{-1} = (k_{nr} + k_r)$. Taking into account the $r^{-6}$ distance dependence for dipole–dipole interactions, the distance dependence of the FRET rate is expressed as a function of the donor lifetime $\tau_D$ and $R_0$ in Eq. (4.5):

$$k_{FRET} = \tau_D^{-1}\left(\frac{R_0}{r_{DA}}\right)^6 \quad (4.5)$$

The FRET rate describes how "fast" the energy transfer occurs. The probability for FRET is described by the FRET efficiency, $E_{FRET}$, which is defined as the ratio of $k_{FRET}$ over all donor decay rates, as shown in the first part of Eq. (4.6).

$$E_{FRET} = \frac{k_{FRET}}{k_r + k_{nr} + k_{FRET}} = \frac{k_{FRET}}{\tau_D^{-1} + k_{FRET}} = 1 - \frac{\tau_{DA}}{\tau_D} \quad (4.6)$$

The FRET efficiency can be determined experimentally by measuring the photoluminescence lifetime of the donor alone, $\tau_D$, and in the presence of the acceptor, $\tau_{DA}$, as indicated in the last part of Eq. (4.6). Including Eq. (4.5) in the definition of $E_{FRET}$, one obtains Eq. (4.7), which allows for the calculation of the expected FRET efficiency from $R_0$ and the donor–acceptor separation $r_{DA}$ for single donor–acceptor systems.

$$E_{FRET} = \frac{R_0^6}{R_0^6 + r_{DA}^6} \quad (4.7)$$

One can see that for $r_{DA} = R_0$, $E_{FRET} = 50\%$. Furthermore, if $R_0$ is known, Eq. (4.7) can be used to measure distances on the nanoscale with FRET [39,40]. Equations (4.5) to (4.7) only hold for single donor–acceptor pairs. If there are several acceptors, numbered $i$, in the proximity of a donor, $k_{FRET}$ corresponds to the sum of all individual FRET rates, $k_i$, describing the energy transfer of this one donor to the $i^{th}$ acceptor at a distance $r_i$, as given in Eq. (4.8).

$$k_{\text{FRET}} = \sum_i k_i = \tau_D^{-1} \sum_i \left(\frac{R_0}{r_i}\right)^6 \tag{4.8}$$

Including this expression for the FRET rate in Eq. (4.6) results in Eq. (4.9), and it can be seen that unless all donor–acceptor separations $r_i$ are known, the FRET efficiency in multi-acceptor systems cannot be estimated from $R_0$.

$$E_{\text{FRET}} = \frac{1}{1+\left[\sum_i \left(\frac{R_0}{r_i}\right)^6\right]^{-1}} \tag{4.9}$$

### 4.2.4 FRET in Two Dimensions

Wolber and Hudson applied Förster's original theory to a two-dimensional structure to obtain a description of FRET in planar systems with random donor and acceptor distributions [41]. They also took into account exclusion zones with radius $R_{\text{ex}}$ around the donors in their calculations. Originally intended for application to biological systems with minimum donor–acceptor separations, this exclusion zone can be used to take into account the nanometre-sized diameter of the QDs, as will be shown in Section 4.3.2.

The time-dependent donor intensity $I_{\text{DA}}(t)$ as a function of $R_0$, $R_{\text{ex}}$ and the acceptor concentration $c_{\text{Acc}}$, as obtained by Wolber and Hudson, is given in Eq. (4.10). $I_{\text{DA}}(t)$ depends on the donor lifetime $\tau_D$ and includes the incomplete Gamma function $\gamma(x,y)\int_0^y z^{x-1}e^{-z}dz$.

$$I_{\text{DA}}(t) = \exp\left\{\begin{array}{l} -\dfrac{t}{\tau_D} - \pi \cdot R_0^2 \cdot c_{\text{Acc}} \cdot \gamma(2/3, (R_0/R_{\text{ex}})^6 \cdot (t/\tau_D)) \cdot \left(\dfrac{t}{\tau_D}\right)^{1/3} \\ +\pi \cdot R_{\text{ex}}^2 \cdot c_{\text{Acc}} \cdot \left[1 - e^{-(R_0/R_{\text{ex}})^6 \cdot (t/\tau_D)}\right] \end{array}\right\} \tag{4.10}$$

The quantum yield $q_r = \int I_{\text{DA}}dt \big/ \int I_D dt$ can be calculated by numerically integrating Eq. (4.10), resulting in Eq. (4.11). It is equivalent to the lifetime ratio $\tau_{\text{DA}}/\tau_D$ and can be used to determine the FRET efficiency, $E_{\text{FRET}} = 1 - q_r$.

$$q_r = \int_0^\infty e^{-\lambda} \exp\left[-\pi C \gamma(2/3, \lambda/r)\lambda^{1/3}\right] \exp\left[\pi C r^{1/3}\left(1 - e^{-\lambda/r}\right)\right] d\lambda \tag{4.11}$$

In Eq. (4.11), $C = R_0^2\, c_{\text{Acc}}$ represents a dimensionless concentration and $r = R_{\text{ex}}/R_0$, corresponding to $r = \alpha^{1/6}$ in the original work by Wolber and Hudson [41]. The values of $R_0$ and $R_{\text{ex}}$ for a two-dimensional donor/acceptor FRET system can be determined by fitting Eqs. (4.10) and (4.11) to the donor photoluminescence decay as well as the concentration dependence of the lifetime ratio, $q_r = \tau_{\text{DA}}/\tau_{\text{D}}$. These two equations can also be used to evaluate energy transfer in structures with two separated, but parallel layers of donors and acceptors.

### 4.2.5 Separated Donor–Acceptor Layer Structure

For a separated donor–acceptor layer structure with a homogeneous acceptor distribution, analytical expressions for the FRET rate and efficiency can also be obtained. Replacing the sum in Eq. (4.8) with an integration over the acceptor plane leads to Eq. (4.12). It can be seen that the interaction of a donor with a plane of acceptors shows a $R_{\text{ex}}^{-4}$ distance dependence of $k_{\text{FRET}}$ instead of the typical $R_{\text{ex}}^{-6}$ distance dependence for a single donor–acceptor pair.

$$k_{\text{FRET}} = \frac{c_{\text{Acc}}\, \pi R_0^6}{2 R_{ex}^4 \tau_{\text{D}}} \tag{4.12}$$

In Eq. (4.12), $R_{\text{ex}}$ represents the separation of the donor and acceptor planes, the minimum distance between a donor and an acceptor. The FRET efficiency for this system is calculated by including Eq. (4.12) in Eq. (4.6), resulting in Eq. (4.13).

$$E_{\text{FRET}} = \frac{1}{1 + \dfrac{2R_{ex}^4}{c_{\text{Acc}}\, \pi R_0^6}} \tag{4.13}$$

In the separated donor–acceptor layer structure, $E_{\text{FRET}}$ depends on $R_0$, $R_{\text{ex}}$ and the acceptor concentration $c_{\text{Acc}}$, but not the donor concentration $c_{\text{Don}}$.

## 4.3 FRET in QD Nanoclusters and QD Solids

### 4.3.1 Monodispersed CdTe QD Molecules and Nanowires

Energy transfer in nanoclusters provides the opportunity to study the energy transfer mechanism on the single donor or acceptor scale. CdTe QD nanoclusters have been formed by covalent and electrostatic interaction [42–44]. Koole *et al.* reported electronic coupling and energy transfer in the smallest QD assembled units that can be formed, namely QD molecules comprising just two, three and four QDs [42]. Allyamine (AA) capped CdTe QDs cross-linked with 1,6 hexanedithiol (HdT) were used to form homonuclear QD molecules using three different sized QDs, green, yellow and red emitting. The fraction of QDs present as single QDs or in QD molecules of different sizes was measured using cryo-TEM and was varied by adjusting the QD:linker ratio.

In the case of the larger red-emitting QDs, the emission decreased and the emission peak red-shifted with increasing QD:linker ratio, as shown in Fig. 4.2a. This can be explained by exciton energy transfer within the molecule, which was further supported by the observation of an initial fast decay in time resolved photoluminescence (PL) data at the ratios where the decreased intensity and red-shifted emission peak occur, Fig. 4.2b. In the case of QD molecules formed using the smaller green-emitting QDs at the same QD:linker ratios, there is evidence for electronic coupling, for example in the changes observed in the absorption spectra which was absent in the case of the larger red-emitting QDs. Stronger electronic coupling for smaller sized QDs can be expected due to greater extension of the wave function outside the QD. In green QD molecules formed using QD:linker ratios less than 1:8, there is no evidence for energy transfer as the carriers are delocalized over the molecule. However, for the larger molecules of more than two QDs formed at higher QD: linker ratios, energy transfer can occur within the molecule as the delocalization does not extend over the entire molecule. Carrier delocalization effects are more important for QDs with diameters less than 2 nm and interparticle separations less than 1 nm [45–47].

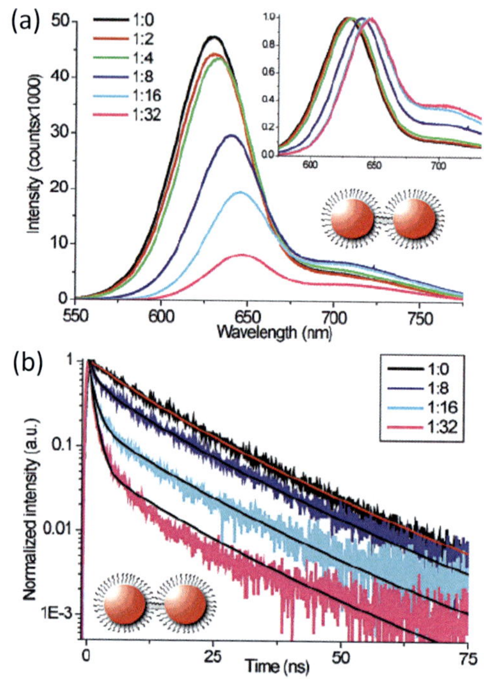


**Figure 4.2** (a) Emission spectra and (b) the photoluminescence decay curves of cross-linked red emitting QDs in toluene at various QD:linker ratios. The solid curve in (b) show the results of fitting with a bi-exponential with the initial fast decay and subsequent slower decay attributed to the QDs that are part of a molecule and the single QDs, respectively. Reproduced with permission from ref. [42]. Copyright 2006, American Chemical Society.

CdTe nanocrystal QDs have also been organized into assemblies of nanowires or nanochains [48,49]. Tang *et al.* reported on the formation of CdTe nanoparticle chains using QDs with a mean size of 3.7 nm and an interparticle separation of 0.5 nm determined by the thioglycolic acid (TGA) ligand [49]. Figure 4.3a shows the TEM images of the chains. The absorption spectra of the chains were identical to the QDs in solution but the PL spectra showed a strong red-shift of the PL emission peak and a reduction of the PL full width at half maximum in comparison with the solution spectra, Fig. 4.3b. The time-resolved PL decays are shown in Fig. 4.3c and it can be observed that both the fast and slow decay lifetimes for the nanoparticle chain are reduced relative to the lifetimes for nanoparticles in solution.

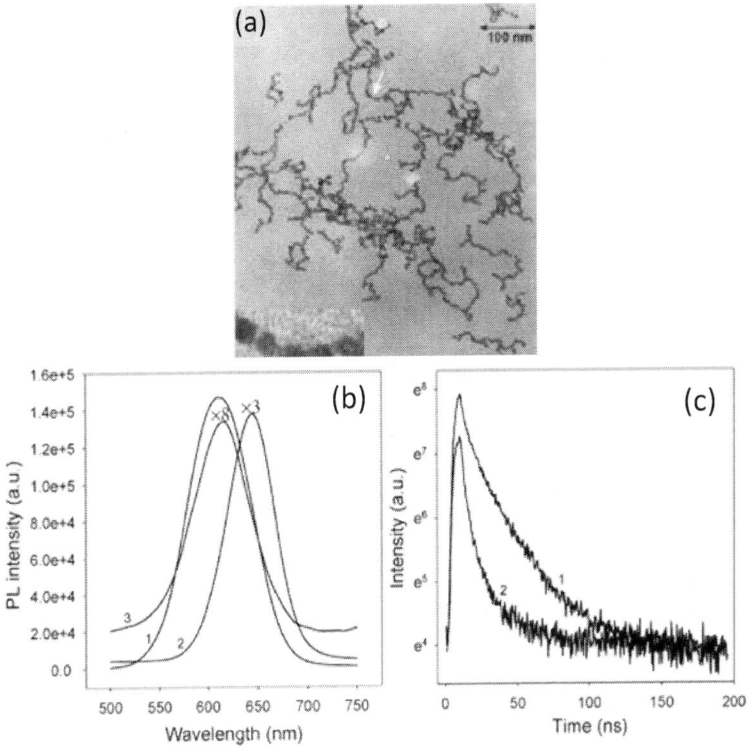


**Figure 4.3** (a) TEM image of CdTe QD chains. The inset shows the enlarged mage of the part of the chain marked by the arrow. (b) PL spectra for free CdTe QDs (1), QD chains (2) and ultrasonically broken chains (3). (c) Time-resolved PL of decay curves of free CdTe QDs (1) and QD chains (2). Reproduced with permission from ref. [49]. Copyright 2004, American Chemical Society.

These spectroscopic signatures, similar to those reported for the red QD molecules above, are consistent with FRET from the smaller higher energy QDs to the larger energy accepting QDs. The authors compared the energy transfer with previously reported results in closely packed solids of CdSe QDs. The energy transfer efficiency, as measured from the change in PL lifetime, is relatively high with a decrease in the PL fast decay component from 7.4 ns for free QDs to 4.3 ns for the nanochain. This is despite the substantial reduction in number of close neighbours for each QD in the chain geometry. Strong FRET in the CdTe QD chain is attributed to the short core to core separation due to the short TGA ligand length and the 10–15% size distribution within the ensemble which is thought to increase the probability for a QD to be sufficiently close to a suitable larger energy accepting QD. There is also the possibility of dipole alignment along the chain which would also increase the FRET efficiency *via* the orientation factor, $\kappa$, discussed in Section 4.2.2. To control the direction of the energy flow size graded QD structures can be formed and energy transferred to the lowest energy nanoparticle *via* cascaded FRET. Unidirectional energy funnelling has been demonstrated in nanowire [48] and multi-layer geometries [19,50]. Energy transfer between QDs of distinctly different sizes will be discussed in Section 4.4.

### 4.3.2 Monodispersed CdTe QD Monolayers

FRET has been extensively reported in nanocrystal QD solids comprised one or more different sized QDs [5,6,9,30,34,51–53]. The three-dimensional solids can be assembled using a variety of techniques including spin-coating, drop-casting, the Langmuir–Blodgett technique and layer-by-layer (LbL) assembly as reviewed in Chapter 3. Many of the structures are multilayered with a monolayer of QDs forming the fundamental building block. Energy transfer within a monodispersed monolayer is expected and it is of interest to understand how strongly energy transfer within the layer influences the optical properties of the monolayer. QD concentration is a key parameter that can be tailored to optimize FRET-based devices and has an impact on the energy transfer efficiency. The QD concentration dependence of FRET in monodispersed NC QD monolayers was investigated for a range of negatively charged, green-emitting colloidal CdTe QDs, stabilized by thioglycolic acid in a aqueous solution [36]. The monolayers were prepared using the LbL technique which had previously been shown to be suitable for the deposition of monolayers of charged colloidal QDs such as CdTe QDs [19,50,54]. This technique is popular due to its relative simplicity. Layered structures of oppositely charged elements can be deposited on substrates with different shapes and morphologies. In particular, as the LbL technique allows for the positioning of layers in nanometre steps, it is well suited to the study of energy transfer in bi-layer or multilayer graded NC QD structures, which will be discussed in more detail later in Section 4.4.

The concentration $c_{QD}$ in the QD monolayers was varied between $0.4 \times 10^{17}$ m$^{-2}$, corresponding to a QD separation of approximately 5 nm, and $3.5 \times 10^{17}$ m$^{-2}$, where closely packed layers are formed. The concentration dependent optical properties were characterized by measurement of the PL spectra and time-resolved PL in three different wavelength ranges across the PL spectrum. Ten different green-emitting QDs were studied with a range of ensemble properties as characterized by the peak emission wavelength, the full width at half maximum (FWHM) of the ensemble PL spectrum, the Stokes' shift between the first absorption peak and the peak emission peak wavelength, measured in solution. The absorption and emission spectra can be used to calculate the spectral overlap, $J$, as given by Eq. (4.3) in Section 4.2.2, which in the case of a monodispersed sample can be referred to as the self-overlap. Red-shifting of the peak emission wavelength, shortening of the photoluminescence lifetime on the high energy side of the ensemble emission spectrum as well as increased difference in the lifetime on the high- and low-energy sides is observed with increasing QD concentration. The QDs with the largest self-overlap showed the greatest changes over the QD concentration range studied, whereas for QDs with the smaller self-overlap these characteristics are almost concentration independent. The lifetime difference is defined as $1 - \tau_{blue}/\tau_{red}$, where the lifetimes $\tau_{blue}$ and $\tau_{red}$ have been determined for the decays on the blue (higher energy) and red (lower energy) side of the PL emission spectrum for all QD layers. It is noted that that while $\tau_{red}$ increases slightly, the influence of the QD concentration is greater on $\tau_{blue}$. The decrease of the lifetime on the blue side of the QD ensemble emission with increasing QD concentration in conjunction with the red-shift of the peak emission is consistent with non-radiative energy transfer from the smaller QDs, emitting on the blue side of the ensemble spectrum, to the larger QDs, emitting on the red side of the ensemble spectrum. As discussed in Section 4.2, since the smallest separation between QDs in the monolayer is 2.7–3.2 nm for the range of QDs studied, the energy transfer can be attributed to the FRET mechanism.

To confirm that the origin of the concentration dependent optical properties of the monodispersed QD monolayers is primarily due to intra-ensemble FRET, the experimentally measured concentration dependence of the lifetime data is compared with that predicted by a theory for FRET in a two-dimensional plane with a random distribution of donors and acceptors [41], as described in Section 4.2.4. The lifetime yield, defined as the ratio of the concentration dependent lifetime and the undisturbed value in the absence of energy transfer, $q_r = \tau(c_{QD})/\tau^0$, is calculated using Eq. (4.11). The theoretical fits to the experimental data are shown in Fig. 4.4a for three representative QD samples.

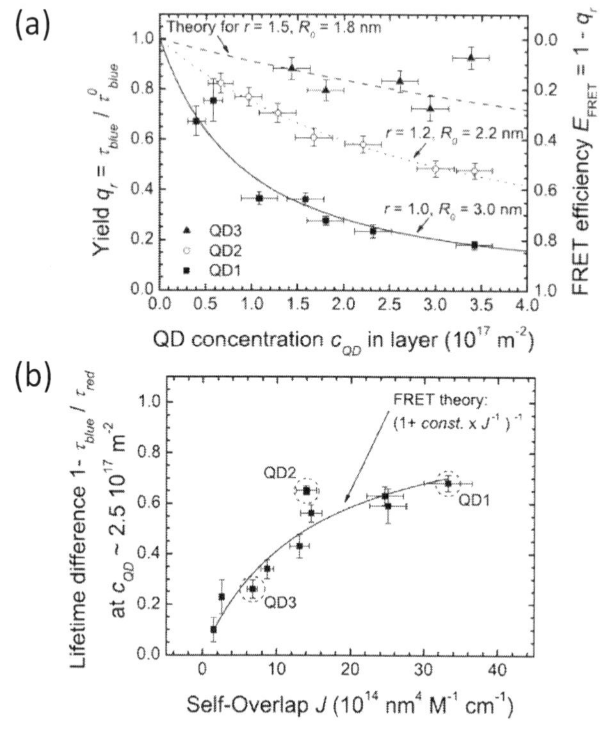


**Figure 4.4** (a) Measured (symbols) and calculated (lines) lifetime yield $q_r$ (left hand axis) as a function of the QD concentration $c_{QD}$ for 3 representative samples QD1 (squares), QD2 (open circles) and QD3 (triangles). The corresponding values of the FRET efficiency $E_{FRET}$ can be read from the right-hand axis. (b) Lifetime difference between the blue and red sides of the QD ensemble emission (squares) for monolayers with a concentration $c_{QD} \approx 2.5 \times 10^{17}$ m$^{-2}$, as a function of the self-overlap $J$ for 10 different QD samples. The theoretically calculated trend (line) is based on the dependence of the FRET efficiency on the self-overlap $J$. Reproduced with permission from ref. [36]. Copyright (2010) by the American Physical Society.

The fit parameters are the exclusion radius, $R_{ex}$, the Förster radius, $R_0$ and $\tau_0$. It can be noted that the exclusion radius, $R_{ex}$, is determined by the size of the QDs which can be extracted from the absorption spectrum. In fitting the lifetime yield $R_{ex}$ was kept within the experimental error of the value determined from the absorption spectrum. The Förster radius, $R_0$, can be independently calculated from the absorption and emission spectra of the QDs (Eq. (4.4) in Section 4.2.2). The good agreement between the experimentally measured lifetime yield and that calculated theoretically with parameters that have been validated independently from the QD spectra measurements confirms that FRET is the dominant mechanism influencing the concentration dependence of the PL lifetime in QD monolayers.

The analysis of the lifetime difference as a function of the self-overlap, shown in Fig. 4.4b, indicates that the lifetime difference can be used to indirectly evaluate the intra-donor FRET efficiency. The solid line represents a fit to the data using an expression for the FRET efficiency of the form of Eq. (4.9)

$$E_{\mathrm{FRET}} = \frac{1}{1+\left[\sum_i (R_0/r_i)^6\right]^{-1}} = \frac{1}{1+\mathrm{const.}\cdot J^{-1}} \tag{4.14}$$

The lifetime difference, which can be measured at a single concentration, is a simpler quantity to determine than the FRET efficiency which requires measurements at more than one concentration in order to extract the undisturbed lifetime $\tau^0_{\mathrm{blue}}$.

## 4.4 FRET in Donor–Acceptor QD Structures

### 4.4.1 FRET in Donor–Acceptor QD Nanoclusters

Nanoclusters formed using two different sized QDs have also been studied [43,44,48,55–57] with evidence of FRET observed in the PL emission in all cases. FRET in donor–acceptor cluster solutions is more efficient than that observed in monodispersed solutions due to the greater availability of suitable acceptors. Mayilo *et al.* reported FRET in clusters formed in pure QD solutions and mixed solutions containing two and three different sized QDs. Cascaded energy transfer was observed in the clusters formed using CdTe QDs emitting in the green, red and near infra-red regions of the spectrum [43]. Higgins *et al.* reported on energy transfer in nanoclusters formed using two different sized oppositely charged CdTe QDs in aqueous solution [55]. Due to the influence of intra-donor ensemble FRET the donor:acceptor ratio dependence of the donor photoluminescence lifetime indicates different regimes of cluster formation including donor–acceptor clusters as well as all-donor clusters. The negatively charged TGA stabilized green-emitting donor QDs and positively charged cysteamine stabilized red-emitting acceptor QDs were mixed in aqueous solution. Maximum quenching of the donor PL was observed for a ratio of four donors per acceptor after which the donor emission recovers slightly before reaching a plateau, Fig. 4.5a. The acceptor PL shows the opposite trend, reaching the largest enhancement at a donor:acceptor ratio of 4:1, Fig. 4.5a. Further increasing the donor:acceptor ratio does not yield any further enhancement of the acceptor PL suggesting that the maximum number of donors that can be attached to a single acceptor has been achieved. This is in good agreement with the zeta potential measurements of +60 meV and –16 meV for the positively charged acceptor and negatively charged donor QDs, respectively. The average donor lifetime, in Fig. 4.5b, shows a similar dependence on the donor:acceptor ratio as the donor PL quenching. The donor concentration independent donor lifetime in the range up to 1:1 indicates that for these ratios every donor is attached to a single acceptor. It can be noted that in the range 1:1 to 5:1 the donor lifetime decreases. As discussed in Section 4.2 the energy transfer rate from donor to acceptor is dependent on the acceptor concentration but is expected to be independent of the number of donors. Given the higher surface charge of the acceptor QDs, the formation of clusters containing more than one acceptor is very unlikely. The decrease of the average donor lifetime with increasing donor:acceptor ratio in the range 1:1 to 5:1 is attributed to donor–donor intra-ensemble energy transfer which becomes possible as multiple donors attach to a single acceptor bringing the donors into closer proximity. Energy transfer within monodispersed QDs has already been discussed above for CdTe QD molecules formed in solution from as few as two QDs and nanowires, Section 4.3.1. The decreasing average donor lifetime as

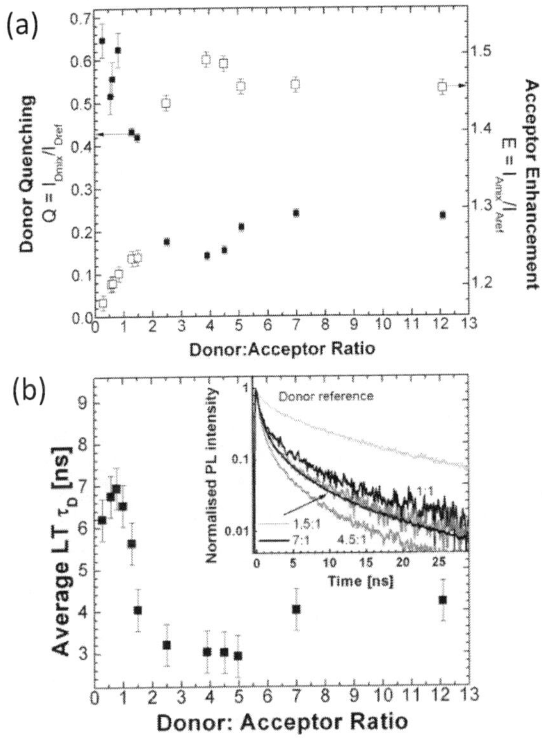


**Figure 4.5** (a) PL quenching of the donor QD emission (left axis) and PL enhancement of the acceptor QD emission (right axis) for mixed solutions at various donor:acceptor ratios. (b) The average donor lifetime for mixed solutions at various donor: acceptor ratios. The measured donor PL decays are shown in the inset. Reproduced with permission from ref. [55]. Copyright 2010, The Optical Society.

the donor:acceptor ratio is increased to 5:1 is therefore a signature of the formation of larger nanoclusters comprising multiple donor QDs attaching to a single acceptor QD. It is also interesting to note that as the donor concentration in the mixed solution is further increased, the average donor lifetime begins to increase, consistent with the PL quenching and acceptor PL enhancement data which indicated that beyond a ratio of 5:1 no additional donors attach to the acceptor QDs. Adding more donors to the mixed solution should increase the number of free donors resulting in an increase of the measured average donor lifetime. However, as can be seen, the measured donor lifetime does not continue to increase but instead levels off. This is attributed to the formation of all-donor clusters in the presence of excess cysteamine in the mixed donor–acceptor QD solution and is supported by the observation of reduced donor lifetimes when cysteamine was added to a pure donor reference solution [55].

### 4.4.2 FRET in Mixed Donor–Acceptor Monolayers

As outlined in the theory section, it is difficult to estimate the FRET efficiency of a multi-acceptor system with random distribution *via* the Förster radius. However, in order to optimize the efficiency and the rate of a FRET system, it is important to understand the link between the different parameters influencing the Förster radius, efficiency and rate.

As discussed in Section 4.3.2 the FRET efficiency as a function of QD concentration in a monolayer structure can be described by a two-dimensional FRET theory taking into account exclusion zones around the donor QDs [41]. This theory can also be applied to analyse FRET in mixed donor–acceptor QD monolayer structures [35]. As shown in Fig. 4.6a, the theory can be used to fit the donor emission decay in the presence of the acceptor QDs (measured data: open circles, fit with theory: dashed line) based on the decay measured in a pure donor QD layer with the same concentration (measurement: solid circles, two-exponential decay fit: solid line). The fitting procedure can be used to extract numbers for the exclusion zone radius $R_{ex}$, the Förster radius $R_0$ as well as the acceptor QD concentration $c_{Acc}$ from the time-resolved donor photoluminescence decays. Furthermore, information on $R_0$ and $R_{ex}$ can also be obtained by analysing the acceptor concentration dependence of the FRET efficiency, similar to the analysis of the lifetime yield in the monodispersed monolayer discussed above in Section 4.3.2 and in reference [37].

The values of $R_0 = (3.6 \pm 0.3)$ nm and $R_{ex} = (2.9 \pm 0.3)$ nm, obtained from fitting the time-resolved donor photoluminescence decays (Fig. 4.6a), in structures with different acceptor QD concentrations discussed in reference [35], agree well with those calculated from the spectral overlap and inferred from the absorption spectra $R_0 = (3.3 \pm 0.2)$ nm and $R_{ex} = (3.2 \pm 0.3)$ nm, respectively. Good agreement of the measured acceptor concentration dependence of the FRET efficiency (Fig. 4.6b) with the theoretically calculated trend for $R_0$ = 3.6 nm and $R_{ex}$ = 2.9 nm is also found, for these mixed monolayer structures.

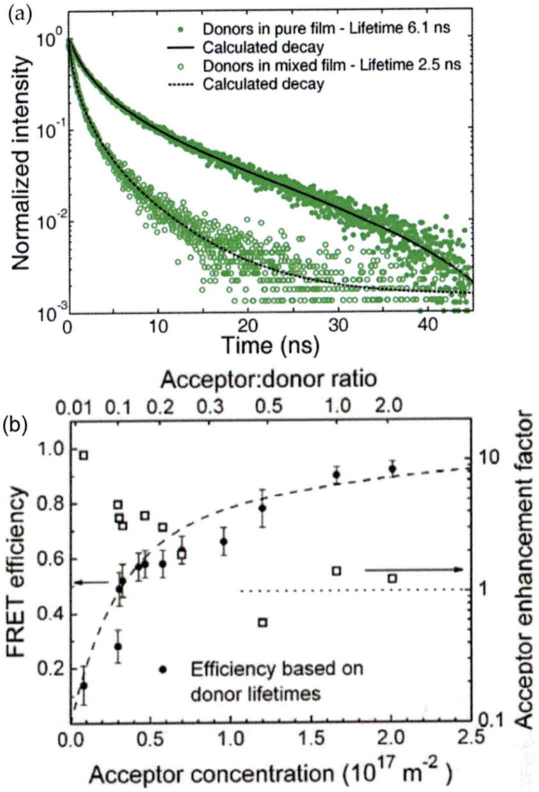


**Figure 4.6** (a) Donor QD decay in a pure donor monolayer (solid green circles) and in a mixed donor–acceptor monolayer (open green circles) with the same donor QD concentration in both layers. The lines represent two-exponential fits to the measured decays. (b) Acceptor concentration dependence of the FRET efficiency (black circles) and acceptor emission increase (white squares). The dashed line is the calculated FRET efficiency as a function of the acceptor concentration. Figure (a) is reproduced with permission from ref. [35]. Copyright 2009, American Chemical Society and figure (b) is reproduced with permission from ref. [53]. Copyright 2010, Elsevier.

This shows that the spectral overlap of the donor emission spectrum with the full acceptor QD absorption spectrum can be used to estimate the Förster radius of a donor QD-acceptor QD FRET system. Furthermore, this also highlights again that the basic FRET theory can successfully describe FRET in QD structures, if the size of the QDs is taken into account in form of exclusion zones in the calculations. Hence, it is possible to calculate all important FRET parameters from donor photoluminescence lifetime measurements only—avoiding the comparison of emission and absorption measurements acquired with different equipment. Further proof for the validity of applying FRET theory to QD systems, was found for a mixed CdSe QD film deposited on an elastomeric film [58]. Investigating the FRET efficiency while stretching the film showed good agreement with the trend of the distance dependence expected from FRET theory.

Analysing the acceptor emission increase, shown in Fig. 4.6c, it can be seen that the acceptor enhancement keeps increasing up to the highest acceptor:donor ratio of 1:33 investigated [35]. This indicates that also for this closely packed mixed CdTe QD monolayer, energy can be transferred from up to the third shell of donors to a single, central acceptor. This finding is in line with the results presented by Achermann *et al.* for closely packed CdSe QDs [51].

### 4.4.3 FRET in Donor–Acceptor Multilayer Structures

The first FRET rate reported in a QD structure was of the order of $(10\ \text{ns})^{-1}$ in a mixed CdSe QD solid, demonstrated by Kagan *et al.* [34]. Due to the strong distance dependence of the FRET process, it is beneficial to reduce the donor–acceptor separation to achieve fast FRET rates and consequently high FRET efficiencies. Therefore, different research groups have presented work on FRET in closely packed and ordered—bilayer—donor–acceptor QD structures [30,51]. Using core-only CdTe QDs and no or just small linker molecules reduces the donor–acceptor separation further, so that even higher FRET rates can be achieved [52,59]. So far, the fastest FRET rate of $(50\ \text{ps})^{-1}$ was reported for a bilayer structure directly assembled from oppositely charged donor and acceptor CdTe QDs [59]. Figure 4.7 shows (a) the time evolution of the emission spectrum of this structure on the nanosecond timescale as well as (b) the donor emission decays for this fast FRET structure.

As mentioned previously, multilayer QD structures are relatively easy to prepare and due to the possibility for fast FRET processes, they have been proposed for different applications. A few examples for such functional CdTe QD multilayer structures are assemblies for white light emission [50], colour tuning [60,61], sensing [62] and energy funnelling structures [19,48].

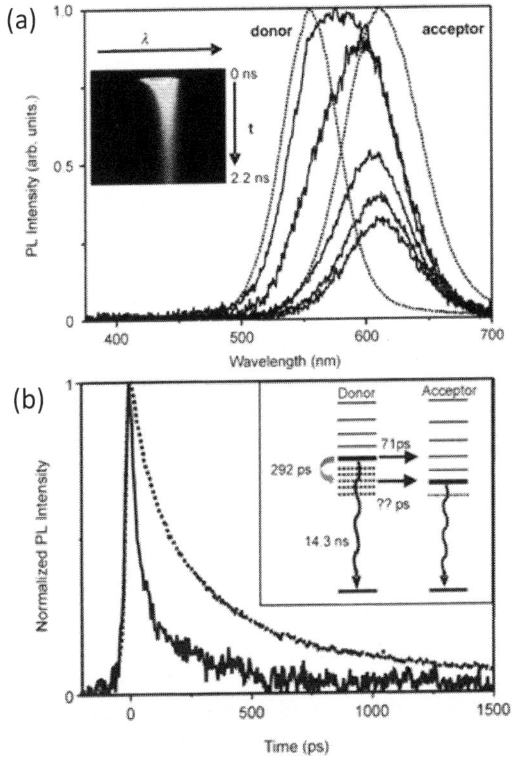


**Figure 4.7** (a) Streak camera image (inset) and photoluminescence spectra of a donor–acceptor CdTe bilayer structure, with minimal layer separation, taken at different times on the nanosecond timescale. (b) Donor photoluminescence decays for a donor monolayer and in a donor–acceptor bilayer structure. The inset illustrates the energy level scheme of the donors and acceptors as well as the decay times for different donor exciton recombination processes. Reproduced with permission from ref. [59]. Copyright 2005, WILEY VCH Verlag GmbH & Co. KGaA.

Moreover, ordered structures with separated donor and acceptor QD layers also offer technical advantages for FRET studies with respect to mixed donor–acceptor QD monolayers or solids. First of all, the distance dependence of the FRET process can be investigated more reliably as the inter-layer spacing and therefore the actual donor–acceptor separation can be tuned and determined more accurately. Furthermore, the three important parameters influencing FRET efficiency and acceptor enhancement:

- acceptor–donor QD separation
- donor QD concentration and
- acceptor QD concentration

can be controlled independently, which is not possible in mixed donor–acceptor QD assemblies, discussed in the previous sections.

From Eqs. (4.12) and (4.13) it can be seen that the donor–acceptor layer separation has a strong influence on the FRET rate and efficiency in a bilayer structure. The typical $R_{ex}^{-4}$ trend of the FRET distance dependence in a QD bilayer structure has first been demonstrated for the FRET rate measured in a CdSe QD system [6]. The experimental results, Fig. 4.8a, for a CdTe QD bilayer structure show that for this system the measured FRET efficiency as a function of the donor–acceptor layer separation also agrees well with the $R_{ex}^{-4}$ distance dependence predicted by theory [37]. Furthermore, a good quantitative agreement with the values of the FRET efficiency calculated with Eq. (4.13) for $R_0$ = 3.9 nm, determined from the spectral overlap, is obtained for the distance dependence measured at low donor concentrations [37,53].

Due to the changes in FRET efficiency for different layer separations, the apparent colour of the overall emission also changes. This variation in the dominant, visible colour is shown in the colour diagram in Fig. 4.8b. The separation dependence of FRET has been proposed as a method to tune colour chromaticity in nanocrystal QD LEDs [60].

In addition to the donor–acceptor separation, Eqs. (4.12) and (4.13) show that the FRET process strongly depends on the acceptor concentration. As can be seen in Fig. 4.9a, the FRET efficiency increases with increasing acceptor QD concentration (symbols) as expected from theory (lines). However, good quantitative agreement with theory calculated for the Förster radii obtained from the spectral overlap at the time of measurement, $R_0$ = 3.9 nm and $R_0$ = 3.1 nm for the measurements at low and high donor concentrations, respectively, is only obtained for the structures with high donor QD concentrations. In order to be able to fit the theory expression to the measured FRET efficiencies at lower donor QD concentrations, an increased $R_0$ = 4.3 nm instead of $R_0$ = 3.9 nm, has to be used.

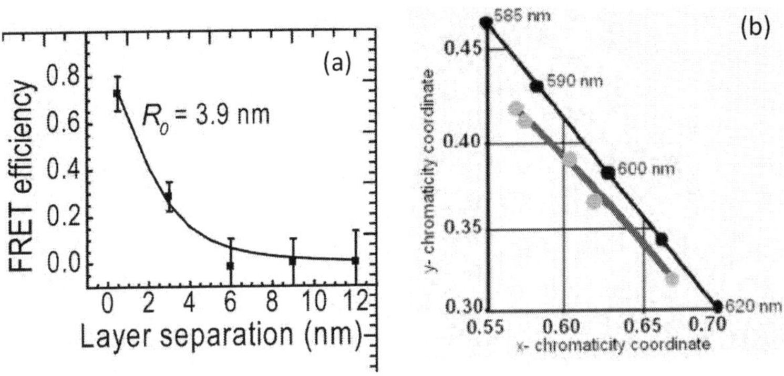


**Figure 4.8** (a) Distance dependence of the FRET efficiency in a CdTe bilayer structure (squares). The line indicates the calculation of the FRET efficiency with basic FRET theory for a Förster radius of 3.9 nm, determined from the spectral overlap of the donor emission and acceptor absorption (reproduced with permission from ref. [37]. Copyright (2011) by the American Physical Society). (b) Changes in the color coordinates of the emission of a CdTe donor–acceptor multilayer structures with varying donor–acceptor layer separation (reproduced with permission from ref. [60]. Copyright (2009) by the American Institute of Physics).

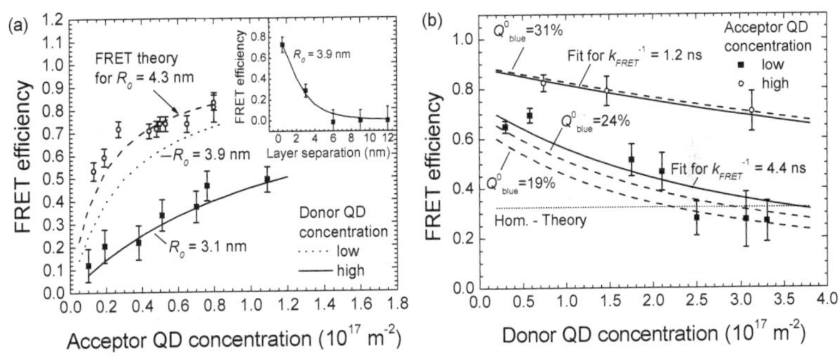


**Figure 4.9** (a) FRET efficiency as a function of the acceptor QD concentration for two different donor QD concentrations. The lines indicate the trends of the FRET efficiency calculated with FRET theory. (b) Donor QD concentration dependence of the FRET efficiency for two different acceptor QD concentrations. The solid lines indicate the result from a fit of the measured data with a FRET rate model and the dashed lines indicate the calculation of the FRET efficiency as a function of the donor QD concentration including a concentration dependent donor quantum yield and different initial, undisturbed quantum yield $Q^0_{\text{blue}}$. Reproduced with permission from ref. [37]. Copyright (2011) by the American Physical Society.

From theory, no impact of the donor concentration on the FRET efficiency from donors to acceptors would be expected. However, due to the inhomogeneous broadening of the donor QD ensemble, energy can be transferred *via* FRET from the blue to the red side of the donor ensemble emission spectrum, as reported in reference [36] and discussed specifically for monolayer geometries in Section 4.3.2. Donor intra-ensemble energy transfer gives rise to a concentration dependent donor lifetime on the blue side of the donor ensemble emission spectrum [36] which also affects the donor quantum yield and the effective Förster radius [37]. The effect of the competition between donor intra-ensemble and donor–acceptor inter-layer FRET manifests in the discrepancy between measured FRET efficiency and theory, taking into account $R_0$ = 3.9 nm, at low donor concentrations in Fig. 4.9a. This effect can also be seen very clearly in the decrease of the FRET efficiency with increasing donor concentration presented in Fig. 4.9b.

In Fig. 4.9b, the FRET efficiency is shown as a function of the donor QD concentration for two different acceptor concentrations. In both cases the FRET efficiency clearly decreases with increasing donor QD concentration. This trend can be well reproduced by a FRET rate model, given in Eq. (4.6), if the concentration dependent donor lifetime measured on the blue side of the donor ensemble emission spectrum, as discussed in references [36] and [37], is included. The FRET rates, extracted from fitting the FRET efficiencies with the rate model, of $(1.2\ \text{ns})^{-1}$ and $(4.4\ \text{ns})^{-1}$, agree well with those determined for other colloidal QD systems [5,6,30,34,51,52,59]. In order to also include the effect of the donor intra-ensemble energy transfer in the calculation of the FRET efficiency *via* the Förster radius with Eq. (4.13), it has been shown that a donor concentration dependent quantum yield has to be included in the calculation of $R_0$ [37]. It is necessary to include the effect of the blue shift in the donor emission spectrum, arising from intra-ensemble energy transfer, as the FRET efficiency is calculated from the donor lifetime measured on the blue side of the donor ensemble emission spectrum. Good agreement of the calculated FRET efficiency with that determined from the donor photoluminescence lifetime measurements is obtained for initial, undisturbed donor quantum yields of approximately 30% (at very low donor concentrations) [37], values similar to those measured for the donor QDs in solution. LbL assembled structures of TGA stabilized CdTe QDs have proven to be an excellent system in which to study the dependences of FRET on separation as well as donor and acceptor concentrations. Quantitative comparison of theory and experiment has indicated the regimes where the competition between intra-ensemble FRET and donor–acceptor FRET must be taken into account in the analysis.

As mentioned previously FRET can be used to engineer energy flow for light harvesting applications. Exciton funnelling *via* cascaded energy transfer was demonstrated in a multilayer QD structure comprised of monolayers of size graded CdTe QDs. Excitons were transferred from outer monolayers of smaller QDs towards a central monolayer of the largest QDs, Fig. 4.10. It was also reported that the quantum efficiency of the cascaded FRET structure is greater than expected from the quantum yields of the individual monolayers and is derived from highly efficient recycling of trapped electron-hole pairs [19]. The exciton recycling efficiency is dependent on the donor–acceptor concentration ratio which can be adjusted to optimise the quantum efficiency of the nanocrystal emission in an LED [63].

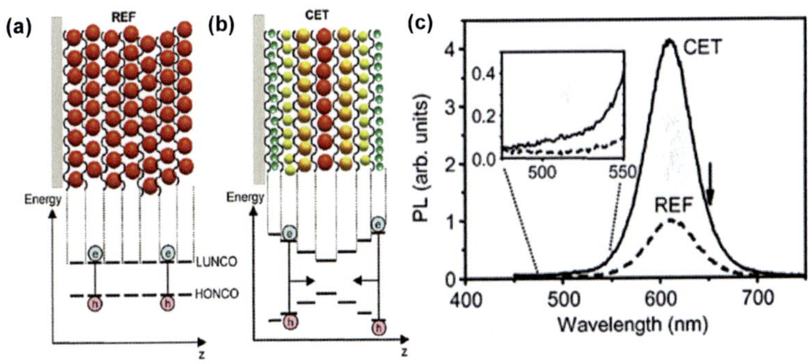


**Figure 4.10** (a) Schematic sketch of the reference sample (REF) which consists of seven layers of red emitting, 3.5 nm large CdTe nanocrystals. (b) Schematic sketch of the cascaded energy transfer sample (CET) consisting of subsequent layers comprising green, yellow, orange, red, orange, yellow, and green emitting nanocrystals. (c) Photoluminescence (PL) of the samples REF and CET. Reproduced with permission from ref. [19]. Copyright 2004, American Chemical Society.

### 4.4.4 Localized Surface Plasmon Enhanced FRET in CdTe Donor–Acceptor Multilayers

The strong distance dependence of FRET can be a limiting factor for some applications or device designs that would benefit from longer interaction distances. The possibility of surface plasmon enhancement of FRET has been both theoretically [64–66] and experimentally investigated in organic [67–69] and QD systems [70–72]. The first report with CdTe QDs considered a mixed donor–acceptor monolayer on top of a layer of gold nanoparticles and evidence for a 2.7-fold increase of the FRET efficiency was observed [70]. The mixed monolayer structure does not allow for independent control of the separation of the donor and acceptor monolayers from the gold nanoparticle layer or the donor and acceptor concentrations.

A later study compared a donor–acceptor bilayer structure with a donor-gold-acceptor sandwich structure with a donor–acceptor centre-to-centre distance of 23.4 nm. In Fig. 4.11, it can be seen that the gold nanoparticle layer quenches the acceptor monolayer emission, with a recovery of the emission only observed after the donor monolayer is added to complete the sandwich structure [71]. The change in the fast component of the donor monolayer lifetime can be used to determine the surface plasmon enhanced FRET rate, $k_F = (6.5 \pm 2.3\ \text{ns})^{-1}$, and efficiency, $E_F = (8 \pm 2)\%$. A corresponding Forster radius of $R_0 = (7.9 \pm 0.5)$ nm can be calculated by rearranging Eq. (4.12). This Förster radius can be compared with that calculated from the measured spectral overlap for this donor–acceptor pair using Eq. (4.4). The FRET efficiency in the bilayer structure (without gold nanoparticles) is too low to be experimentally measured at this large donor–acceptor separation but Eqs. (4.12) and (4.13) could be used to calculate the expected FRET rate, $k_F = (517 \pm 230\ \text{ns})^{-1}$ and FRET efficiency, $E_F = (0.12 \pm 0.05)\%$. A comparison of the measured and calculated values for the bilayer and sandwich structure shows a 103% increase in the Förster radius and an 80-fold increase in the energy transfer rate due to the interaction with localized surface plasmons in the sandwich structure.

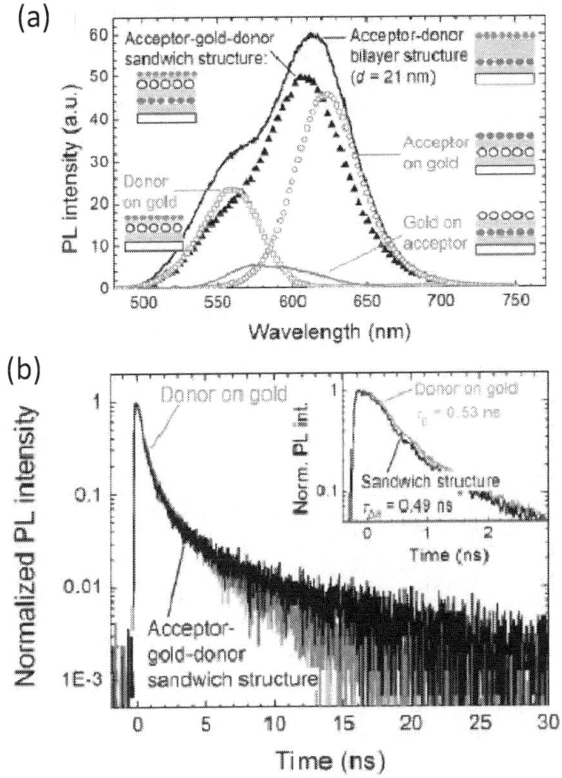


**Figure 4.11** (a) PL spectra of the acceptor–gold–donor, acceptor on gold, donor on gold and acceptor–donor structures. (b) The donor PL decays for the donor on gold and acceptor–donor–gold structures. The inset shows the two decays over the first three nanoseconds. Reproduced with permission from ref. [71]. Copyright 2011, American Chemical Society.

## 4.5 Summary and Outlook

All-QD structures are an excellent laboratory for the study of FRET and as discussed in this chapter, FRET has been observed in nanocrystal CdTe QD assemblies of all dimensions. The absence of additional shells and the short ligand chains make TGA stabilized CdTe QDs interesting building blocks for FRET structures. Quantitative comparison of theory and experiment has shown that

QD FRET systems can be well described within the standard models developed for dipole–dipole interaction.

Due to the inhomogeneous broadening of QD ensembles, FRET also occurs within monodispersed QD assemblies and this intra-QD ensemble energy transfer is seen to introduce significant changes in the optical properties of QD assemblies relative to those of the QDs in solution. The competition between intra-donor ensemble FRET and donor–acceptor FRET can introduce a donor concentration dependence of the measured FRET efficiency, which can be included in the standard FRET equations in a simple way. Furthermore, surface plasmon enhancement of the FRET rate and energy transfer distance has also been demonstrated for CdTe QD donor–acceptor systems.

FRET in both all-QD and hybrid-QD systems can form the basis of many applications. Given the large variety of assembly techniques, geometries and materials that can be combined with nanocrystal QDs there are many opportunities yet to be explored. Improved control and understanding of FRET with QDs will lead to greater optimization and flexibility of new device designs. In Chapter 5, other aspects of FRET in hybrid CdTe-biological systems are presented.

## References

1. Rogach, A. L. (2008). *Semiconductor Nanocrystal Quantum Dots*, (Springer, Wien).
2. Talapin, D. V., Lee, J. S., Kovalenko, M. V., and Shevchenko, E. V. (2010). Prospects of colloidal nanocrystals for electronic and optoelectronic applications, *Chem. Rev.*, **110**, pp. 389–458.
3. Shavel, A., Gaponik, N., and Eychmüller, A. (2005). The assembling of semiconductor nanocrystals, *Eur. J. Inorg. Chem.*, **2005**, pp. 3613–3623.
4. Allan, G., and Delerue, C. (2007). Energy transfer between semiconductor nanocrystals: validity of Förster's theory, *Phys. Rev. B*, **75**, 195311.
5. Kagan, C. R., Murray, C. B., and Bawendi, M. G. (1996). Long range resonance transfer of electronic excitations in close packed CdSe quantum dot solids, *Phys. Rev. B*, **54**, pp. 8633–8643.
6. Kim, D., Okahara, S., Nakayama, M., and Shim, Y. (2008). Experimental verification of Förster energy transfer between semiconductor quantum dots, *Phys. Rev. B*, **78**, 153301.

7. Kim, D., Okazaki, K., and Nakayama, M. (2009). Temperature dependence of the energy transfer of exciton states in bilayer structures of CdSe/ZnS quantum dots, *Phys. Rev. B*, **80**, 045322.

8. Scholes, G. D., and Andrews, D. L. (2005). Resonance energy transfer and quantum dots, *Phys. Rev. B*, **72**, 125331.

9. Wuister, S. F., Koole, R., de Mello Donega, C., and Meijerink, A. (2005). Temperature-dependent energy transfer in cadmium telluride quantum dot solids, *J. Phys. Chem. B*, **109**, pp. 5504–5508.

10. Medintz, I. L., Clapp, A. R., Mattoussi, H., Goldman, E. R., Fisher B., and Mauro, J. M. (2003). Self-assembled nanoscale biosensors based on quantum dot FRET donors, *Nat. Mater.*, **2**, pp. 630–638.

11. Medintz, I. L., and Mattoussi, H. (2009). Quantum dot-based resonance energy transfer and its growing application in biology, *Phys. Chem. Chem. Phys.*, **11**, pp. 17–45.

12. Snee, P. T., Somers, R. C., Nair, G., Zimmer, J. P., Bawendi, M. G., and Nocera, D. G. (2006). A ratiometric CdSe/ZnS nanocrystal pH sensor, *J. Am. Chem. Soc.*, **128**, pp. 13320–13321.

13. Willard, D. M., Mutschler, T., Yu, M., Jung, J., and Van Order, A. (2006). Directing energy flow through quantum dots: toward nanoscale sensing, *Anal. Bioanal. Chem.*, **384**, pp. 564–571.

14. Achermann, M., Petruska, M. A., Kos, S., Smith, D. L., Koleske, D. D., and Klimov, V. I. (2004). Energy-transfer pumping of semiconductor nanocrystals using an epitaxial quantum well, *Nature*, **429**, pp. 642–646.

15. Achermann, M., Petruska, M. A., Koleske, D. D., Crawford, M. H., and Klimov, V. I. (2006). Nanocrystal-based light-emitting diodes utilizing high-efficiency nonradiative energy transfer for colour conversion, *Nano Lett.*, **6**, pp. 1396–1400.

16. Chanyawadee, S., Lagoudakis, P. G., Hartley, R. T., Charlton, M. D. B., Talapin, D. V., Huang, H. W., and Lin, C.-H. (2010). Increased colour-conversion efficiency in hybrid light-emitting diodes utilizing non-radiative energy transfer, *Adv. Mat.*, **22**, pp. 602–606.

17. Nizamoglu, S., and Demir, H. V. (2008). Resonant nonradiative energy transfer in CdSe/ZnS core/shell nanocrystal solids enhances hybrid white light emitting diodes, *Opt. Express*, **16**, pp. 13961–13968.

18. Chanyawadee, S., Harley, R. T., Henini, M., Talapin, D. V., and Lagoudakis, P. G. (2009). Photocurrent enhancement in hybrid nanocrystal quantum-dot p-i-n photovoltaic devices, *Phys. Rev. Lett.*, **102**, 077402.

19. Franzl, T., Klar, T. A., Schietinger, S., Rogach, A. L., and Feldmann, J. (2004). Exciton recycling in graded gap nanocrystal structures, *Nano Lett.*, **4**, pp. 1599–1603.
20. Lu, S., and Madhukar, A. (2007). Nonradiative resonant excitation transfer from nanocrystal quantum dots to adjacent quantum channels, *Nano Lett.*, **7**, pp. 3443–3451.
21. Mutlugun, E., Samarskaya, O., Ozel, T., Cicek, N., Gaponik, N., Eychmüller, A., and Demir, H. V. (2010). Highly efficient nonradiative energy transfer mediated light harvesting in water using aqueous CdTe QD antennas, *Opt. Express*, **18**, pp. 10720–10730.
22. Nabiev, I., Rakovich, A., Sukhanova, A., Lukashev, E., Zagidullin, V., Pachenko, V., Rakovich, Y. P., Donegan, J. F., Rubin, A. B., and Govorov, A. O. (2010). Fluorescent quantum dots as artificial antennas for enhanced light harvesting and energy transfer to photosynthetic reaction centres, *Angew. Chem. Int. Ed.*, **49**, pp. 7217–7221.
23. Rogach, A. L. (2011). Fluorescence energy transfer in hybrid structures of semiconductor nanocrystals, *Nano Today*, **6**, pp. 355–365.
24. Förster, T. (1948). Zwischenmolekulare Energiewanderung und Fluoreszenz, *Annalen Der Physik*, **2**, pp. 55–75.
25. Dexter, D. L. (1953). A theory of sensitized luminescence in solids, *J. Chem. Phys.*, **21**, pp. 836–850.
26. Inokuti, M., and Hirayama, F. (1965). Influence of energy transfer by exchange mechanism on donor luminescence, *J. Chem. Phys.*, **43**, pp. 1978–1989.
27. Lakowicz, J. R. (1999). *Principles of Fluorescence Spectroscopy*, Kluwer Academic/Plenum: New York ; London, 2nd edition.
28. Van der Meer, B. W., Coker, G., and Chen, S. Y. S. (1994). *Resonance Energy Transfer: Theory and Data*, VCH: New York; Cambridge.
29. Faure, S., Stern, C., Guilard, R., and Harvey, P. D. (2004). Role of the spacer in the singlet-singlet energy transfer mechanism (Förster vs Dexter) in cofacial bisporphyrins, *J. Am. Chem. Soc.*, **126**, pp. 1253–1261.
30. Crooker, S. A., Hollingsworth, J. A., Tretiak, S., Klimov, V. I. (2002). Spectrally resolved dynamics of energy transfer in quantum-dot assemblies: Towards engineered energy flows in artificial materials, *Phys. Rev. Lett.*, **89**, 186802.
31. Baer, R., and Rabani, E. (2008). Theory of resonance energy transfer involving nanocrystals: the role of high multipoles, *J. Chem. Phys.*, **128**, 184710.

32. Curutchet, C., Franceschetti, A., Zunger, A., and Scholes, G. D. (2008). Examining Förster energy transfer for semiconductor nanocrystalline quantum dot donors and acceptors, *J. Phys. Chem. C*, **112**, pp. 13336–13341.

33. Schrier, J., and Wang, L. W. (2008). Shape dependence of resonant energy transfer between semiconductor nanocrystals, *J. Phys. Chem. C*, **112**, pp. 11158–11161.

34. Kagan, C. R., Murray, C. B., Nirmal, M., and Bawendi, M. G. (1996). Electronic energy transfer in CdSe quantum dot solids, *Phys. Rev Lett.*, **76**, pp. 1517–1520.

35. Lunz, M., Bradley, A. L., Chen, W. Y., and Gun'ko, Y. K. (2009). Two-dimensional Förster resonant energy transfer in a mixed quantum dot monolayer: experiment and theory, *J. Phys. Chem. C*, **113**, pp. 3084–3088.

36. Lunz, M., Bradley, A. L., Chen, W.-Y., Gerard, V. A., Byrne, S. J., Gun'ko, Y., Lesnyak, V., and Gaponik, N. (2010). Influence of quantum dot concentration on Förster resonant energy transfer in monodispersed nanocrystal quantum dot monolayers, *Phys. Rev. B*, **81**, 205316.

37. Lunz, M., Bradley, A. L., Gerard, V. A., Byrne, S. J., Gun'ko, Y. K., Lesnyak, V., and Gaponik, N. (2011). Concentration dependence of Förster resonant energy transfer between donor and acceptor nanocrystal quantum dot layers: Effect of donor–donor interactions, *Phys. Rev. B*, **83**, 115423.

38. Borst, J. W., Laptonek, S. P., Westphal, A. H., Kühnemuth, R., Hornen, H., Visser, N. V., Kalinin, S., Aker, J., van Hoek, A., Seidel, C. A. M., and Visser, A. J. W. G. (2008). Structural changes of yellow cameleon domains observed by quantitative FRET analysis and polarized fluorescence correlation spectroscopy, *Biophys. J.*, **95**, pp. 5399–5411.

39. Stryer, L., and Haugland, R. P. (1967). Energy transfer—a spectroscopic ruler, *PNAS*, **58**, pp. 719–726.

40. Stryer, L. (1978). Fluorescence energy-transfer as a spectroscopic ruler, *Ann. Rev. Biochem.*, **47**, pp. 819–846.

41. Wolber, P. K., and Hudson, B. S. (1979). Analytic solution to the Förster energy-transfer problem in 2 dimensions, *Biophys. J.*, **28**, pp. 197–210.

42. Koole, R., Liljeroth, P., De Mello Donegá, C., Vanmaekelbergh, D., and Meijerink, A. (2006). Electronic coupling and exciton energy transfer in CdTe quantum-dot molecules, *J. Am. Chem. Soc.*, **128**, pp. 10436–10441.

43. Mayilo, S., Hilhorst, J., Susha, A. S., Höhl, C., Franzl, T., Klar, T. A., Rogach, A. L., and Feldmann, J. (2008). Energy transfer in solution-based clusters of CdTe nanocrystals electrostatically bound by calcium ions, *J. Phys. Chem. C*, **112**, pp. 14589–14594.

44. Osovsky, R., Shavel, A., Amirav, L., Eychmüller, A., Weller, H., and Lifshitz, L. (2005). Electrostatic and covalent interactions in CdTe nanocrystalline assemblies, *J. Phys. Chem. B*, **109**, pp. 20244–20250.

45. Artemyev, M. V., Bibik, A. I., Gurinovich, L. I., Gaponenko, S., and Woggon, U. (1999). Evolution from individual to collective electron states in a dense quantum dot ensemble, *Phys. Rev. B*, **60**, pp. 1504–1506.

46. Artemyev, M. V., Woggon, U., Jaschinski, H., Gurinovich, L. I., and Gaponenko, S. (2000). Spectroscopic study of electronic states in an ensemble of close-packed CdSe nanocrystals, *J. Phys. Chem. B*, **104**, pp. 11617–11621.

47. Micic, O. I., Ahrenkiel, S. P., and Nozik, A. J. (2001). Synthesis of extremely small InP quantum dots and electrostatic coupling in their disordered solid films, *Appl. Phys. Lett.*, **78**, pp. 4022–4024.

48. Lee, J., Govorov, A. O., and Kotov, N. (2005). Bioconjugated superstructures of CdTe nanowires and nanoparticles: multistep cascade Förster resonance energy transfer and energy channeling, *Nano Lett.*, **5**, pp. 2063–2069.

49. Tang, Z., Birol, O., Wang, Y., and Kotov, N. A. (2004). Simple preparation strategy and one-dimensional energy transfer in CdTe nanoparticle chains, *J. Phys. Chem. B*, **108**, pp. 6927–6931.

50. Mamedov, A. A., Belov, A., Giersig, M., Mamedova, N. N., and Kotov N. A. (2001). Nanorainbows: Graded semiconductor films from quantum dots, *J. Am. Chem. Soc.*, **123**, pp. 7738–7739.

51. Achermann, M., Petruska, M. A., Crooker, S. A., and Klimov, V. I. (2003). Picosecond energy transfer in quantum dot Langmuir–Blodgett nanoassemblies, *J. Phys. Chem. B*, **107**, pp. 13782–13787.

52. Franzl, T., Koktysh, D. S., Klar, T. A., Rogach, A. L., Feldmann, J., and Gaponik, N. (2004). Fast energy transfer in layer-by-layer assembled CdTe nanocrystal bilayers, *Appl. Phys. Lett.*, **84**, pp. 2904–2906.

53. Lunz, M., Bradley, A. L., Chen, W.-Y., and Gun'ko, Y. K. (2010). Förster resonant energy transfer in quantum dot layers, *Superlattices Microstruct.*, **47**, pp. 98–102.

54. Tang, Z., Wang, Y., and Kotov, N. A. (2002). Semiconductor nanoparticles on solid substrates: film structure, intermolecular interactions, and polyelectrolyte effects, *Langmuir*, **18**, pp. 7035–7040.

55. Higgins, C., Lunz, M., Bradley, A. L., Gerard, V. A., Byrne, S., Gun'ko, Y. K., Lesnyak, V., and Gaponik, N. (2010). Energy transfer in colloidal CdTe quantum dot nanoclusters, *Opt. Express*, **18**, pp. 24486–24494.

56. Ni, T., Nagesha, D. K., Robles, R., Materer, N. F., Mussig, S., and Kotov, N. A. (2002). CdS nanoparticles modified to chalcogen sites: New supramolecular complexes, butterfly bridging, and related optical effects, *J. Am. Chem. Soc.*, **124**, pp. 3980–3992.

57. Wargnier, R., Baranov, A., Masolov, V. G., Stsiapura, V., Artemyev, M., Pluot, M., Sukhanova, A., and Nabiev, I. (2004). Energy transfer in aqueous solutions of oppositely charged CdSe/ZnS core/shell quantum dots and in quantum dot-nanogold assemblies, *Nano Lett.*, **4**, pp. 451–457.

58. Chen, C. W., Wang, C. H., Chen, Y. F., Lai, C. W., and Chou, P. T. (2008). Tunable energy transfer efficiency based on the composite of mixed CdSe quantum dots and elastomeric film, *Appl. Phys. Lett.*, **92**, 051906.

59. Franzl, T., Shavel, A., Rogach, A. L., Gaponik, N., Klar, T. A., Eychmüller, A., and Feldmann, J. (2005). High-rate unidirectional energy transfer in directly assembled CdTe nanocrystal bilayers, *Small*, **1**, pp. 392–395.

60. Cicek, N., Nizamoglu, S., Ozel, O., Mutlugun, E., Karatay, D. U., Lesnyak, V., Otto, T., Gaponik, N., Eychmüller, A., and Demir, H. V. (2009). Structural tuning of color chromaticity through nonradiative energy transfer by interspacing CdTe nanocrytsal monolayer, *Appl. Phys. Lett.*, **94**, 061105.

61. Lin, Y. W., Tseng, W. L., and Chang, H. T. (2006). Using a Layer-by-Layer assembly technique to fabricate multicolored-light-emitting films of CdSe/CdS and CdTe quantum dots, *Adv. Mater.*, **18**, pp. 1381–1386.

62. Seker, U. O. S., Ozel, T., and Demir, H. V. (2011). Peptide-mediated constructs of quantum dot nanocomposites for enzymatic control of nonradiative energy transfer, *Nano Lett.*, **11**, pp. 1530–1539.

63. Nizamoglu, S., Akin, O., and Demir, H. V. (2009). Quantum efficiency enhancement in nanocrystals using nonradiative energy transfer with optimised donor–acceptor ratio for hybrid LEDs, *Appl. Phys. Lett.*, **94**, 243107.

64. Gersten, J. I., and Nitzan, A. (1984). Accelerated energy-transfer between molecules near a solid particle, *Chem. Phys. Lett.*, **104**, pp. 31–37.

65. Govorov, A. O., Lee, J., and Kotov, N. A. (2007). Theory of plasmon-enhanced Förster energy transfer in optically excited semiconductor and metal nanoparticles, *Phys. Rev. B*, **76**, 125308(1).

66. Hua, X. M., Gersten, J. I., and Nitzan, A. J. (1985). Theory of energy-transfer between molecules near solid-state particles, *Chem. Phys.*, **83**, pp. 3650–3659.

67. Lessard-Viger, M., Rioux, M., Rainville, L., and Boudreau, D. (2009). FRET enhancement in multilayer core-shell nanoparticles, *Nano Lett.*, **9**, pp. 3066–3071.

68. Lessard-Viger, M., Brouard, D., and Boudreau, D. J. (2011). Plasmon-enhanced resonance energy transfer from a conjugated polymer to fluorescent multilayer core-shell nanoparticles: a photophysical study, *Phys. Chem. C*, **115**, pp. 2974–2981.

69. Zhang, J., Fu, Y., and Lakowicz, J. R. (2007). Enhanced Förster resonance energy transfer (FRET) on a single metal particle, *J. Phys. Chem. C*, **111**, pp. 50–56.

70. Komarala, V. K., Bradley, A. L., Rakovich, Y. P., Byrne, S. J., Gun'ko, Y. K., and Rogach, A. L (2008). Surface plasmon enhanced Förster resonance energy transfer between CdTe quantum dots, *Appl. Phys. Lett.*, **93**, 123102.

71. Lunz, M., Gerard, V. A., Gun'ko, Y. K., Lesnyak, N., Gaponik, N., Susha, A. S., Rogach, A. L., and Bradley, A. L. (2011). Surface plasmon enhanced energy transfer between donor and acceptor CdTe nanocrystal quantum dot monolayers, *Nano Lett.*, **11**, pp. 3341–3345.

72. Wang, C. H., Chen, C. W., Chen, Y. T., Wei, C. M., Chen, Y. F., Lai, C. W., Ho, M. L., Chou, P. T., and Hofmann, M. (2010). Surface plasmon enhanced energy transfer between type I CdSe/ZnS and type II CdSe/ZnTe quantum dots, *Appl. Phys. Lett.*, **96**, 071906.

# Chapter 5

# Emission of CdTe Nanocrystals Coupled to Microcavities

**Yury P. Rakovich**[a,b] **and John F. Donegan**[c,d]

[a]*Centro de Fisica de Materiales (CSIC-UPV/EHU), Donostia International Physics Center, P Manuel de Lardizabal 4-5, Donostia-San Sebastian, 20018, Spain*
[b]*IKERBASQUE, Basque Foundation for Science, Alameda Urquijo 36-5, Bilbao, 48011, Spain*
[c]*School of Physics, Trinity College Dublin, Dublin 2, Ireland*
[d]*CRANN Research Institute, Trinity College Dublin, Ireland*

yury.rakovich@ehu.es, jdonegan@tcd.ie

## 5.1 Introduction to Spherical Microcavities

Glass or polymer microspheres, also called "photonic atoms," can act as three-dimensional optical resonators providing the feedback required for the enhancement of linear and non-linear optical processes including photoluminescence (PL), bistability and Raman scattering [1]. The unique spectroscopic properties of spherical particles arise from the confinement of photons within the cavity. The circular optical modes in such microresonators, commonly observed as the ripple structure in elastic scattering efficiencies

*Cadmium Telluride Quantum Dots: Advances and Applications*
Edited by John F. Donegan and Yury P. Rakovich

ISBN 978-981-4316-05-7 (Hardcover), 978-981-4316-09-5 (eBook)
www.panstanford.com

is now referred to by a number of names: cavity resonances, Mie-resonances, morphology dependent resonances or most frequently dubbed as whispering gallery modes (WGMs). The last name came from famous St. Paul's cathedral in London, whose acoustic phenomenon was investigated by Lord Rayleigh [2].

In the optics domain, WGMs can be understood as internal standing waves, with an integer number of wavelengths circulating near the sphere surface (Fig. 5.1a) or, in terms of geometric optics, as closed circular beams supported by total internal reflections from boundaries of the resonators (Fig. 5.1b).

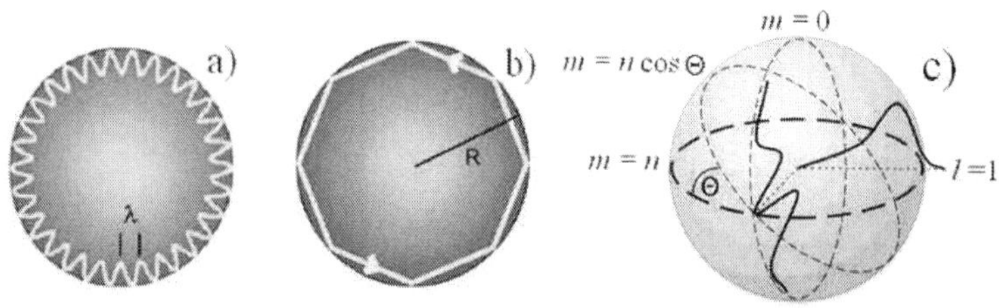


**Figure 5.1** Light confinement inside a microsphere in a wave optics description (a) and in geometric optics picture (b). Schematic diagram of a spherical microcavity, showing the distribution and the spatial orientation of modes (c).

In the absence of gain, the WGM resonances can be characterized by a mode number $n$ (angular quantum number), mode order $l$ (radial quantum number), and azimuthal mode number $m$ (azimuthal quantum number). The value of $n$ is proportional to the circumference divided by the wavelength of the light propagating within the microsphere, the mode order, $l$, indicates the number of maxima in the radial distribution of the internal electric field, and the azimuthal mode number, $m$, gives the orientation of the WGMs orbital plane. The modes offering the highest spatial photon confinement correspond to high values of angular momentum, $n$, smaller values of $l$, and to azimuthal numbers quantum numbers $m$ close to $n$.

The electromagnetic fields confined in a microsphere are given by [3]

$$\mathbf{E}_{\mathrm{TE}} = j_n(nkr)\,\mathbf{X}_{nm}(\theta,\phi), \tag{5.1}$$

for modes having no radial components of the electric field (transverse electric or TE modes) and

$$\mathbf{E}_{\text{TM}} = \nabla \times j_n(nkr)\mathbf{X}_{nm}(\theta,\phi) \tag{5.2}$$

for the transverse magnetic (TM) modes (no radial component of the magnetic field).

Here $\mathbf{X}_{nm}(\theta,\phi)$ are the vector spherical harmonics given in spherical polar coordinates, $j_n(n_r kr)$ and represent the spherical Bessel functions, where $k = 2\pi/\lambda$ is the wavenumber, and $\lambda$ is the free space wavelength.

On the other hand, the wave functions $\Psi_{slm}$ for the electron confined in the hydrogen atom are given by [4]

$$\Psi_{slm} = R_{sl}(r)\mathrm{Y}_{lm}(\theta,\phi), \tag{5.3}$$

where $R_{sl}(r)$ is known as the Laguerre polynomial. The eigen functions (5.1) and (5.3) are very similar and their spatial distributions are characterized by the three integers $s$, $l$, and $m$ (for the hydrogen atom) and $n$, $l$ and $m$ (for the microsphere), which correspond to the total angular, radial and the azimuthal quantum numbers, respectively. Based on these similarities, this approach has enabled small dielectric spheres to be considered "photonic atoms" [1,5]. However, unlike energy states of an electron in the atom, photonic states in spherical microcavities are not localized due to finite storage time of photons in the resonant mode. This "photon lifetime" is controlled by quality factor, $Q$, of the WGMs and therefore can be limited by diffractive losses, absorption, gain, shape deformation or refractive index inhomogeneities. As a result, the resonant internal field of a spherical cavity is not completely confined to the interior of the microparticle and WGMs interact with the environment through their evanescent "tail," which can extend into the cavity exterior by tens or hundreds of nanometers (depending on the sphere size and refractive index contrast).

For spherically symmetric microparticles, the extinction cross section is derivable from the Lorentz–Mie theory[6]:

$$C_{\text{ext}} = \frac{2\pi}{k^2}\mathrm{Re}\sum_{n=1}^{\infty}(2n+1)(b_n(x,n_r) + a_n(x,n_r)), \tag{5.4}$$

where $x = 2\pi R/\lambda$ is the size parameter, $R$ is the radius of microsphere and the Mie scattering partial wave amplitudes $a_n(x,n_r)$ and $b_n(x,n_r)$ can be expressed in the form [7]

$$a_n(x,n_r) = \frac{A_n(x,n_r)}{A_n(x,n_r) + iC_n(x,n_r)}$$

and

$$b_n(x,n_r) = \frac{B_n(x,n_r)}{B_n(x,n_r) + iD_n(x,n_r)}. \tag{5.5}$$

The pairs of functions $A_n$, $C_n$ and $B_n$, $D_n$ are determined by the Mie scattering problem [8] and resonance structure in scattering spectra can be expected when the real part of $a_n$ or $b_n$ reaches its maximum value of 1 and the imaginary part passes through 0 from the positive to the negative side. In other words, the resonances in the Mie scattering characteristics occur when $C_n = 0$ or $D_n = 0$, which gives the following mathematical condition for a resonance:

$$n_r \psi_n(x) \psi_n'(n_r x) - \psi_n(n_r x) \psi_n'(x) = 0 \tag{5.6}$$

or

$$n_r \chi_n(x) \psi_n'(n_r x) - \psi_n(n_r x) \chi_n'(x) = 0 \tag{5.7}$$

Note that for given $n$ and $m$, these equations have infinitely many solutions at discrete values of $x$ for TM and TE WGMs, respectively. Also Eqs. (5.6) and (5.7) are independent of the azimuthal mode number $m$ because of the spherical symmetry implying that the $m$-modes are wavelength degenerate. The Riccati–Bessel functions of the first and second kind can be introduced as follows:

$$\psi_n(z) = z j_n(z) = \sqrt{\frac{\pi z}{2}} J_{n+(1/2)}(z);$$

$$\chi_n(z) = z y_n(z) = \sqrt{\frac{\pi z}{2}} Y_{n+(1/2)}(z); \tag{5.8}$$

where $J(z)$ and $Y(z)$ are the cylindrical Bessel functions of the first and second kind, respectively. The use of Bessel functions for systems with cylindrical symmetry together with reccurence relation $\psi_n'(z) = -\frac{n}{z}\psi_n(z) + \psi_{n-1}(z)$ enables us to reduce Eq. (5.6)–(5.7) to a form convenient for practical calculation of the position of the WGMs.

Thus, if the refractive index is real (no absorption), the TM and TE resonances can be taken in the following form:

$$\frac{1}{x}\left(\frac{n}{n_\mathrm{r}}-n_\mathrm{r}n\right)J_{n+(1/2)}(x)J_{n+(1/2)}(n_\mathrm{r}x)+n_\mathrm{r}J_{n+(1/2)}(n_\mathrm{r}x)J_{n-(1/2)}(x)$$
$$-J_{n+(1/2)}(x)J_{n-(1/2)}(n_\mathrm{r}x)=0, \tag{5.9}$$

$$n_\mathrm{r}Y_{n+(1/2)}(x)J_{n-(1/2)}(n_\mathrm{r}x)-J_{n+(1/2)}(n_\mathrm{r}x)Y_{n-(1/2)}(x)=0. \tag{5.10}$$

These conditions are transcendental equations, which can be solved for the size parameter $x$ (position of a resonance) for given values of refractive index and for given angular quantum number $n$.

Figure 5.2 shows the normalized extinction cross section calculated using Eq. (5.4) for a microsphere with refractive index $n_\mathrm{r}$ = 1.45. Two main features can be seen in this figure: (i) slow oscillations with relatively broad maxima and minima resulting from an interference effect between diffracted and transmitted waves [9] and (ii) sharp resonance peaks superimposed upon this oscillating curve called the ripple structure also known as the set of WGMs [10].

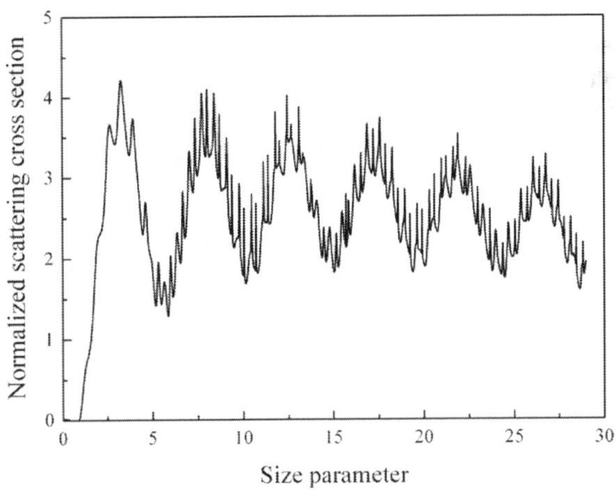


**Figure 5.2** Extinction cross section calculated using Eq. (5.4) for a microsphere with refractive index $n_\mathrm{r}$ = 1.45 shows oscillations with superimposed sharp resonance peaks.

Nowadays, WGMs in spherical microcavities have been a subject of intense theoretical and experimental research paving the way for the development of single-photon emitters, nanojets, sensors, and low-threshold lasers and providing the foundation for fundamental studies of optical matter interactions [11]. The understanding gained from the studies of WGMs cavities gives the possibility of creating new materials and photonic structures with localized states in the optical spectrum. For various applications, the combination of WGMs resonators with highly efficient luminescent materials is especially attractive.

## 5.2 Light Confinement in WGM Microcavities with Semiconductor Nanocrystals

The interest in the design and fabrication of optical structures for the manipulation of light at the micrometer-length scale using semiconductor nanocrystals (NCs) or quantum dots derives from both the fundamental study of the photon confinement with its quantum-electrodynamical consequences and from the potential applications.

Various phenomena can be studied using NCs/microsphere hybrid photonic structures, where both carriers and photons are confined within correspondingly nano- and micron length scales in three dimensions. The basic effect of a microcavity on coupled NCs is expected to be enhancement or suppression of the radiative decay rate. As the spontaneous-emission rate is directly proportional to the fluorescence quantum yield, both can be modified when the emitting species are placed in a small-volume optical enclosure [3].

The fabrication of semiconductor NCs multilayers on spherical microcavities is of immense interest given their tunable and desirable optical and electronic properties for various applications. Among polymers, the high optical transparency and thermal and mechanical stability of melamine formaldehyde (MF) make it interesting as a potential candidate in optical microcavity applications. The refractive index of MF in the visible region ($n_r = 1.68$) is greater than that of polymethylmethacrylate ($n_r = 1.48$), silica ($n_r$ = 1.47) or other glass materials ($n_r$ ~ 1.5). Aqueous dispersions of MF microspheres of various sizes are available for example from Microparticles GmbH (Berlin) and deposition of monolayer NCs shells using layer-by-layer (LbL) electrostatic assembly is well established technique [12] (see also Chapter 3).

The size of the MF spheres used in our experiments (2–4 μm) was dictated by specific requirements for optimal excitation conditions, such as the laser wavelength matching to one of the WGMs frequencies and achieving good correlation between the WGMs and the laser linewidths. Also the wide separation of WGMs in spheres of this size allows us to avoid intricate mode mixing in the PL spectra and apply a well-known theoretical approach to the analysis of mode structure. The film on the surface of microspheres assembled by LbL method contains CdTe NCs at a volume ratio of ~30% as estimated from the initial concentration of nanoparticles in aqueous solution. Assuming that the refractive index is proportional to the volume ratio of the NCs, we calculated its value to be ~2.8.

High-resolution PL confocal microscopy imaging clearly shows the localization of NCs shell at the rim of 3.3 μm MF microsphere (Fig. 5.3, inset). Also, in contrast to the broad, featureless PL band in the spectra of colloidal NCs, the emission spectra of a single MF/NC microsphere exhibit very sharp periodic structure (Fig. 5.3). The observed WGMs peak structure is a result of coupling of electronic states in NCs and photon states of microsphere mediated by exponentially decaying evanescent field of microcavity. Theory suggests that by placing an excited emitter (such as an atom or molecule or NCs) near the surface of a spherical dielectric, the interaction of the emitter with its self-scattered field can produce enormous enhancements in its emission rate at frequencies associated with WGMs [13]. Moreover the refractive index of NCs film is significantly greater than that of MF microspheres. It was recently shown [14] that the deposition of the layer with such a high refractive index onto the microsphere surface can tune the resonances and draw the maximal electromagnetic field outward intensifying the evanescent field and, in our case, providing optimum conditions for efficient coupling of light emitted by CdTe NCs to WGMs of the microcavity.

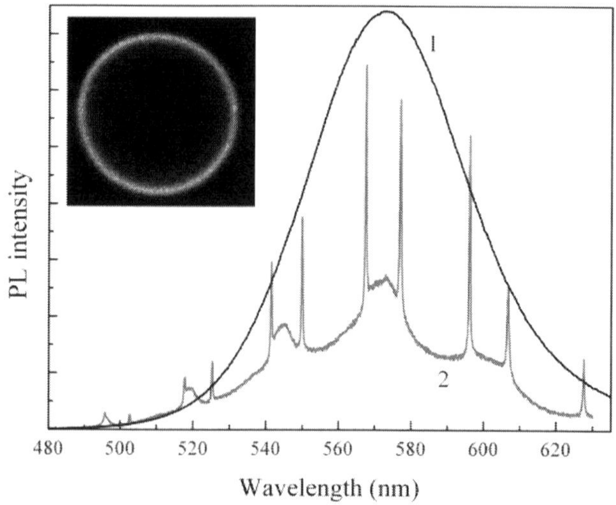


**Figure 5.3** PL spectra of colloidal CdTe NCs in water (1) and micro-PL spectra from a single MF microsphere of 3.3 μm diameter covered by two monolayer shells of CdTe NCs (2). PL spectra of NCs were recorded using a Spex Fluorolog spectrometer. The micro-PL spectra from the single microsphere were recorded using a RENISHAW micro-Raman system. The inset shows confocal PL intensity image of the hybrid MF/NCs microcavity.

Among all the factors that cause a modification in the PL emission pattern due to spatial photon confinement in spherical microcavities, the polarized nature of the WGMs plays an important role because of its potential use in all-optical switches and logical devices [15]. Direct experimental identification of polarization states of WGMs can be performed by inserting a polarizer into the optical beam path in front of the detection system. In this case, the polarizer selects only signals from NCs that emit components of the electromagnetic field parallel to the orientation of the polarizer axis.

For a focal spot on the top rim of a microsphere (Fig. 5.4, inset), the emission spectrum detected with a polarizer orientation perpendicular to the polarization plane of the laser is shown in Fig. 5.4a. Due to the higher excitation efficiency of transverse magnetic modes provided by this experimental scheme [16], the dominant spectral features in this case are WGMs of TM polarization. Rotating the polarizer by 90° leads to a similar WGM structure but with TE peaks dominating the PL spectrum (Fig. 5.4b). This result confirms the polarized nature of WGM resonances; the two adjacent peaks representing TE and TM modes of a microcavity have orthogonal polarization states. The experimental concept of polarization-sensitive mode analysis represents a convenient tool for determining the polarization properties, in particular for larger microspheres, without performing extensive numerical calculations.

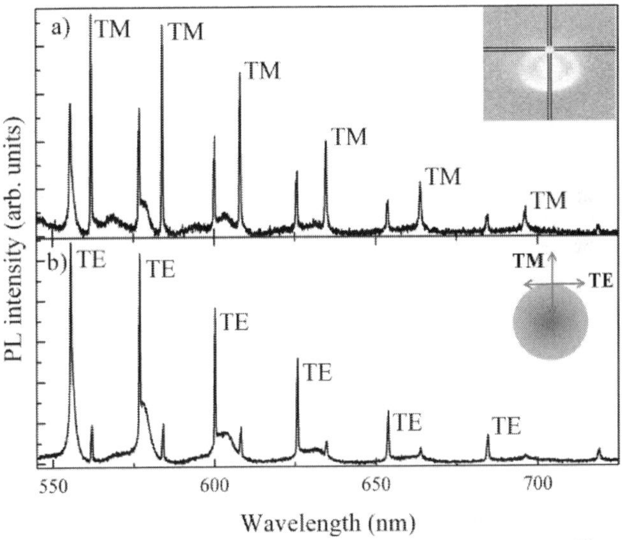


**Figure 5.4** Micro-PL spectra of a single microsphere with a monolayer of CdTe NCs with the polarizer orientation perpendicular to the polarization plane of the $Ar^+$ laser (a), and with the polarizer rotated by 90° (b). The insets show a microscope image of the microcavity with the crosshairs indicating the excitation-detection position (a) and polarization of TE and TM modes with respect to the rim of a microsphere (b).

For small microspheres of 2–10 μm size, the indexes $n$ and $l$ of the WGMs can be identified comparing the calculated results with the spectral positions of the WGM in the experimental photoluminescence (PL) or emission spectra of the microsphere. Additionally, absorptive losses can be accounted for by taking into consideration a small imaginary part of the refraction index.

The algorithm of the mode assignment is as follows. (1) The resonant wavelengths corresponding to the WGM resonances $\lambda_i^{\text{exp}}$

($i = 1, 2, \ldots, N$) are determined from a PL spectrum of a single sphere. (2) We assume an approximate value of the microsphere radius based on the technical specification within distribution of the sizes that are specified by the manufacturer. (3) Theoretical resonance positions $\lambda_i^{\text{theor}}$ are then calculated using Eqs. (5.9) and (5.10). (4) The two lists are compared and for each value of $\lambda_i^{\text{exp}}$ and the closest value $\lambda_i^{\text{theor}}$ is suggested and the difference between them $\Delta_i$ is calculated. (5) Taking into account the spectral resolution $\Delta$, the correlation $S = (1/N)\sum_1^N (1 + \Delta_i / \Delta)^{-1}$ is then maximized by adjusting only one fitting parameter, namely the size of the microsphere. Figures 5.5 and 5.6 show a result of WGM identification in PL spectra of NCs attached to single spherical microcavities of different sizes.

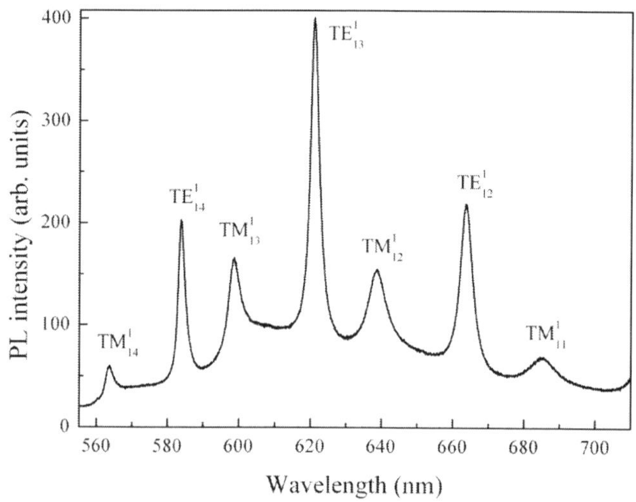


**Figure 5.5** Micro-PL spectra of a single 1.98 μm microsphere with a monolayer of CdTe NCs with modes identified using Eqs. (5.9) and (5.10).

Figures 5.4 and 5.5 also demonstrate two important aspects, the knowledge of which can contribute to the understanding and interpretation of experimental results and to further applications. First, the experimentally observable phenomenon is that the spectral spacing between two resonances of successive modes with the same order and polarization strongly depends on the size of the spherical microcavity. If the refractive index of the microsphere is constant and unchanged during the measurement, this spectral spacing can be expressed in terms of the size parameter as [10]

$$\Delta x = x_{n+1} - x_n \cong \frac{\tan^{-1}(n_r^2 - 1)^{1/2}}{n_r^2 - 1}, \tag{5.11}$$

for resonances with $x_{n,l} >> 1$, $n >> 1$, and $x_{n,l} \sim n$.

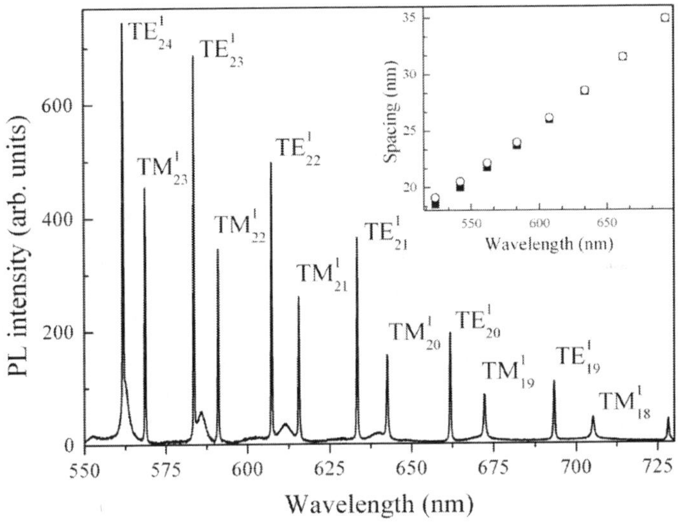


**Figure 5.6** Micro-PL spectra of a single 3 μm microsphere with a monolayer of CdTe NCs with modes identified using Eqs. (5.9) and (5.10). Because of the high PL quantum efficiency of the NCs, the WGMs in the PL spectra are superimposed on a background signal arising from the part of the NCs emission, which does not match any WGMs of the microsphere. This PL background has been subtracted in the spectra presented in this figure to show the WGMs structure more clearly. The inset shows experimental (squares) and calculated (circles) spacing between adjacent TE modes.

In the small sphere limit, when $|x - n| >> 1/2$, the following relation applies [7]

$$\Delta x = x_{n+1} - x_n \cong \frac{x_n \tan^{-1}[(n_r x_n / n)^2 - 1]^{1/2}}{n[(n_r x_n / n)^2 - 1]^{1/2}}. \tag{5.12}$$

According to Eq. (5.12), the mode spacing decreases with increasing mode number $n$. The calculated and experimentally observed $\Delta\lambda$ (or $\Delta x$) values for modes, which differ by one unit of angular momentum, are shown in the inset to Fig. 5.6, demonstrating good agreement and indicating that the monolayer of semiconductor material with a high refractive index does not influence the dielectric response of the microcavity.

Since the experimental resonant spectrum can be precisely fitted using the Mie theory, it was shown that for non-absorbing microspheres the difference between the mode spacing obtained by Eq. (5.11) or Eq. (5.12) and by experiment is ~1% [7]. For example, from the resonant spectrum in Fig. 5.6, the mode spacing between resonances $TE^1_{24}$ and $TE^1_{23}$ measured in terms of wavelength is 21.8 nm, hence the experimental mode spacing can be fitted by the above asymptotic formulas with an accuracy within 1%, i.e., a deviation <0.22 nm, when a suitable sphere size, refractive index and mode number are chosen. This simplified approach is largely exploited in current analysis of WGM structures, especially for large values of the size parameter when the use of the whole machinery of Mie scattering theory algebraically or even computationally is cumbersome.

The second phenomenon clearly seen by comparing Fig. 5.5 and Fig. 5.6 is the significant broadening of WGMs peaks in the spectrum corresponding to microspheres having a size below 10 microns, compared with the spectrum corresponding to bigger microcavities, which reflects the lower quality factor, $Q$, of smaller microspheres. Since WGMs are, in fact, leaky modes, the quality factor of a resonance is defined as [17]

$$Q = \frac{\text{Stored energy}}{\text{Energy lost per cycle}} \tag{5.13}$$

Also, the $Q$ factor is the number of cycles required for the stored energy to decrease to $e^{-\pi}$ times of its original value. A high quality-factor is the basis for the interest in most of the suggested applications for microspheres.

Near a WGM, the electric field of a resonant mode inside the microcavity varies as

$$E(t) = E_0 \exp\left[\left(i\omega_0 - \frac{\omega_0}{2Q}\right)t\right], \tag{5.14}$$

where $\omega_0$ is the resonant frequency. The distribution of the energy of a near-resonant mode in a microcavity $|E(t)|^2$, is therefore proportional to the Lorentzian

$$\frac{1}{(\omega-\omega_0)^2+\left(\frac{\omega_0}{2Q}\right)^2} \qquad (5.15)$$

The spectral width between two points at which the energy is half of its maximum value is

$$\Delta\omega=\frac{\omega_0}{Q}, \qquad (5.16)$$

which also defines the average lifetime of the photon in resonant mode as

$$\tau=\frac{1}{\Delta\omega_0}=\frac{Q}{\omega_0}. \qquad (5.17)$$

Apart from diffractive losses, the actual $Q$ of the microcavity is determined by various factors such as radiative losses and losses or gain attributable to absorption or emission by the material of the cavity, refractive index inhomogeneity or shape deformation. Also, a strong decrease in the value of the $Q$ factor is expected when the radius of curvature of the microsphere is comparable (within an order of magnitude) to the resonant wavelength of the WGM. This effect is accompanied by broadening of the corresponding spectral peaks (Figs. 5.5 and 5.6).

The typical value of the $Q$ factor for composite MF/CdTe NCs microcavity of 3 μm size is ~3500, which corresponds to the value of average photon storage time ~1 ps. Decreasing the size of microsphere down to 2 μm results in a $Q$ value of ~800 and a corresponding modification of the photon storage lifetime.

## 5.3 Modes Manipulation in Single Microcavities with Semiconductor Nanocrystals

Resonant photonic modes in spherical microcavities have high sensitivity to changes of the refractive index of the sphere itself [18] or to the environment in close vicinity to the sphere surface [19,20]. In the last case, the sensitivity derives from the ability of the evanescent field of a WGM to explore the surrounding medium. Along with these properties, WGMs show remarkable sensitivity to deformation, which allows for fine tuning of the WGMs spectral position and $Q$ factor [21,22].

As discussed earlier, WGMs are wavelength degenerate with respect to the azimuthal modes because of the spherical symmetry of the microcavity, which means that orbits with various inclinations, described by various $m$-modes have the same path lengths and consequently the same resonance wavelengths. The total number of these $m$-modes, originating from a certain mode $n$, is governed by Eqs. (5.1) and (5.2) and in the approach of the normal mode concept is $2n + 1$.

One of the ways to reveal different $m$-modes is to deform the microsphere, for example, using a compression device [21]. Shape distortion breaks the azimuthal symmetry and results in a lifting of the degeneracy. In a deformed sphere, the photonic orbits with various inclinations have different path lengths and therefore the spectral positions of $m$-modes are no longer the same. In other words, a single WGM peak in the spectrum of a perfect sphere is now split into a series of azimuthal ($m$) resonances. The $m$-modes with "+" and "−" signs, however, designate clockwise and counter-clockwise rotations, respectively, and therefore still have the same resonance wavelength. As a result, no degeneracy is removed between $m$ and $-m$ in deformed microspheres and the new degeneracy of WGMs is now $n + 1$ with $m$ running from 0 to $n$. Perturbation theory provides an analytical expression for the frequency shifts and mode splitting of each of the WGMs for an oblate spheroid with a small amplitude of distortion $e = (r_\mathrm{p} - r_\mathrm{e})/R \ll 1$, where the polar and equatorial radii are $r_\mathrm{p}$ and $r_\mathrm{e}$, respectively [23]:

$$\lambda(m) = \lambda_{n,l}\left\{1 + \frac{e}{6}\left[1 - \frac{3m^2}{n(n+1)}\right]\right\} \tag{5.18}$$

where $\lambda_{n,l}$ is the resonance wavelength for a perfect sphere.

The distortion amplitude $e$ is defined to be positive for prolate spheroids and negative for oblate spheroids. For an oblate spheroid, the $m = \pm n$ equatorial mode has the lowest resonance frequency (i.e., the longest resonance wavelength) (Fig. 5.7). An $m$-mode of certain $n$ can be envisioned as filling $n$ waveguides around the perimeter of the spheroid along the great circle of the $m$-mode. For an oblate spheroid, it then becomes physically intuitive to associate the smallest perimeter of the polar great circle $m = 0$ with a shorter wavelength and a higher frequency.

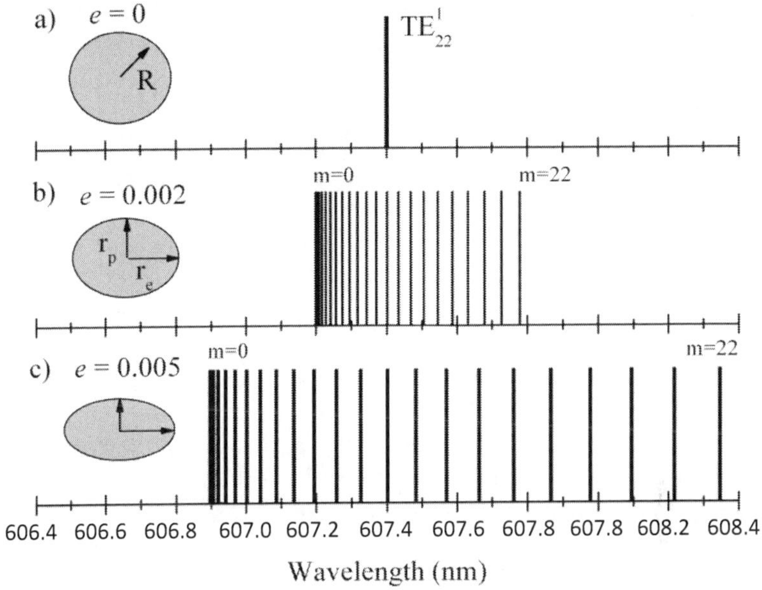


**Figure 5.7** The evolution of the WGM resonance from a perfect sphere (a) to a slightly distorted oblate spheroid (b) and to a severely distorted oblate spheroid (c). In (b) and (c), *m*-modes appear to be merged if spectrally unresolved, or are split if spectrally resolved.

Apart from the mechanical deformation of the microsphere, lifting of WGM degeneracy may be obtained by other means. The long evanescent tail of confined photonic states in a spherical microcavity implies high sensitivity of WGM line shape to the surface quality. In microspheres with a thin surface layer, the evanescent tail of the field induces a dipole moment in surface shell, causing a shift in the energy of the resonant state [14]. This is due to a change in the refractive index at the interface between the microsphere and the deposited layer.

The PE shell deposited on the surface of MF microspheres exhibits transient or permanent pores forming channels from inside to outside [24]. The value of Young's modulus of PE LbL films (100–200 MPa) falls in the range characteristic for highly cross-linked rubber and indicates that LbL film is in fact a physically cross-linked network of PE molecules. Due to strong difference in elastic properties of the MF and PE multiplayer film [25] deformation of PE shell may be the crucial factor controlling the WGM structure in PL spectra of the microcavity. However, controllable mechanical deformation of 3 μm microspheres combined with PL spectroscopy is difficult to realize. The alternative is to use the radiation force acting on a microcavity near a substrate illuminated by a tightly focused laser beam. In this experimental scheme instead of deforming the whole microsphere, only the soft PE layer is compressed under radiation pressure causing modification of the WGM linewidth and structure.

It is well known that the radiation pressure created by the focused laser beam can be used to trap, levitate and manipulate micro- or nano-sized dielectric particles and biological cells [26,27]. Although photons are massless particles, they can transfer their momentum to matter. Let us suppose that a spherical microcavity with a PE/CdTe shell is positioned near a nonconductive homogeneous substrate. We assume the origin of the reference Cartesian coordinate system to be on the substrate surface, and $z$- and $y$-axes are normal and parallel to the substrate surface, respectively (Fig. 5.8). A laser beam of wavelength $\lambda$ is tightly focused with an objective lens of high numerical aperture. The strong gradient of the electromagnetic field intensity in the region of the beam waist gives rise to the gradient force (Fig. 5.8). Depending on the focus position, the gradient force can work against the gravitational force, providing a means for the optical binding and the manipulation of ultra-fine particles and mesoscopic systems. On the other hand, a beam of light can exert sufficient radiation pressure to move a microcavity along the direction of the beam propagation under the effect of a scattering force. The interplay of scattering and gradient forces provides a way of trapping the particle within the beam and manipulation in all three dimensions. Both these forces scale linearly with light intensity and when the gradient force dominates the scattering force, a restoring force with a magnitude of piconewtons acts to confine the particle to the most intense region in the laser beam and a stable trap is created.

For an object of radius $R$ in air and a light beam with wavelength $\lambda$ and intensity $I_0$, the force resulting from the light scattering is [28]

$$F_{\text{scat}} = I_0 n_{\text{m}} \frac{\sigma\langle S\rangle}{c}, \tag{5.19}$$

where $n_m$ is the refractive index of the surrounding medium, $\langle S \rangle$ is the time-averaged Poynting vector, $c$ is the speed of light, and $\sigma$ is the particle's scattering cross-section, which in the case of a spherical particle is given by

$$\sigma = \frac{8}{3}\pi(kR)^4 R^2 \left(\frac{n_r^2 - 1}{n_r^2 + 2}\right)^2 \tag{5.20}$$

with the refractive index of the particle $n_r$ and $k = 2\pi/\lambda$, the wave vector of the light.

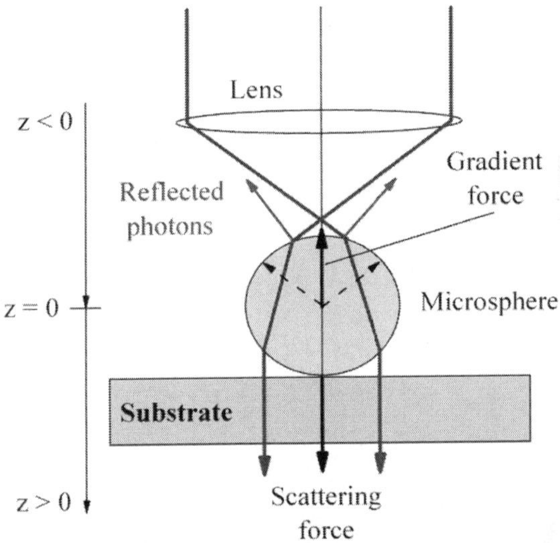


**Figure 5.8** Radiative forces on a spherical microcavity.

The intensity gradient near the beam focus gives rise to a gradient force. This force is due to the Lorentz force acting on the dipole, induced by the electromagnetic field. In the field of the laser, the gradient force is given by [28]

$$F_{\text{grad}} = \frac{-n_m^3 R^3}{2}\left(\frac{n_r^2 - 1}{n_r^2 - 2}\right)\nabla(|E|^2), \tag{5.21}$$

where $E$ is the electric field vector of the laser light. An essential feature predicted by this equation is that the gradient force is always directed toward the region of highest light intensity.

For a microcavity size larger than the wavelength of laser irradiation, the net force exerted on the microsphere by a focused

Gaussian beam can be calculated by the use of generalized Lorentz–Mie theory (GLMT) [29]. The original Lorentz–Mie theory [30] is valid for light scattering of a plane electromagnetic wave from the particle. The GLMT enables us to calculate the spectrum of the scattered light for different positions of the particle in the beam and for different types of incoming waves. In this approach, the radiation pressure exerted by the laser beam propagating toward the positive $z$ direction and impinging on a spherical microparticle located along the axis of beam is expressed by

$$F_{\mathrm{rad}} = \frac{2P}{c\pi\omega_0^2} C_{\mathrm{pr}}, \tag{5.22}$$

where $P$ is the power of the laser beam, $\omega_0$ the spot size of the focused beam and $C_{\mathrm{pr}}$ the radiation pressure cross-section, which is given by [29]

$$C_{\mathrm{pr}} = \frac{\lambda^2}{2\pi} \sum_{n=1}^{\infty} \left\{ \begin{aligned} & \frac{2n+1}{n(n+1)} |g_n|^2 \operatorname{Re}\left(a_n + b_n - 2a_n b_n^*\right) \\ & + \frac{n(n+2)}{n+1} \operatorname{Re}\left[ g_n g_{n+1}^* \left(a_n + b_n + a_{n+1}^* + b_{n+1}^* - 2a_n a_{n+1}^* - 2b_n b_{n+1}^*\right)\right] \end{aligned} \right\} \tag{5.23}$$

In this equation the beam shape coefficient $g_n$ for an on-axis Gaussian beam is given by

$$g_{\mathrm{n}} = \left(1 + i2s\frac{z_0}{\omega_0}\right) \exp(ikz_0) \exp\left[\frac{-s^2(n-1)(n+2)}{1+i2(sz_0/\omega_0)}\right], \tag{5.24}$$

where $s$ is a small parameter defined by $s = 1/k\omega_0$, and $z_0$ is the coordinate of the beam-waist center. The Mie-coefficients $a_n$ and $b_n$ are defined by Eq. (5.5).

Figure 5.9 shows the calculated radiation force $F_{\mathrm{pr}}$ for a variable focus position along the $z$-axis in the center of the sphere with a diameter of 3 μm. A negative value of $F_{\mathrm{pr}}$ means a reverse radiation force exerted on the microsphere, pulling the sphere in the opposite direction of the beam propagation. This happens when the resulting gradient force exceeds the scattering force. A positive $F_{\mathrm{pr}}$ pushes the sphere downwards onto the solid substrate. The value $z$ = 0 on the $x$-axis in Fig. 5.9 corresponds to the center of the sphere (Fig. 5.8). Positive $z$ values are focus positions below the center toward the substrate and negative values of $z$ correspond to a focus point above the sphere center (Fig. 5.8). A freely moving microsphere would be trapped at the point where the radiation force is zero. In these calculations, it is nearly 0.5 μm below the top of the sphere and 1.0 μm above the sphere center, respectively ($z$ = −1 μm). If a laser beam is focused at this point, no radiation force is exerted on the sphere. If the focus is moved further up, the resulting radiation force is pulling the sphere upward and if the focus is moved further down, the sphere is pushed onto the substrate with maximum radiation pressure at around 0.6 μm ($z$ = 0.9 μm) above the substrate. These results clearly demonstrate the possibility to control the radiation pressure by tuning the focus position, implying a non-contact method of controllable tuning modification of WGMs in a spherical microcavity with a compressible surface shell.

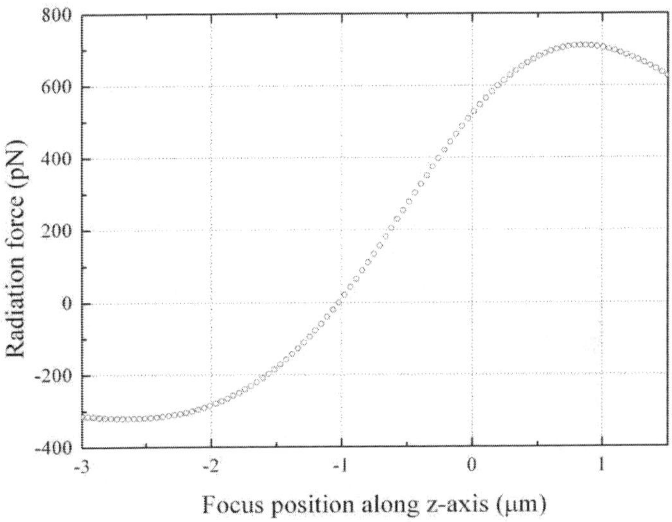


**Figure 5.9** Calculation of radiation pressure applied on a MF microsphere of 3 μm size along the $z$-axis. Excitation power $P$ = 20 mW.

Exploring this possibility, the PL spectra of a single microsphere with 15 nm thick PE shell and a monolayer of CdTe NCs were taken at different focus positions and excitation power (Fig. 5.10). In order to modify the stiffness of the PE shell, an additional chemical procedure was used. Before the spheres are coated with the NC layer, the sample was kept for 3 days in a highly concentrated NaCl (3 Mol/L) solution. This treatment reduces the solidity of the PE layer resulting in a softer surface layer [31]. The soft layer is necessary to irreversibly deform the spherical shape of the PE shell by radiation pressure without mechanical devices being in contact with the sphere.

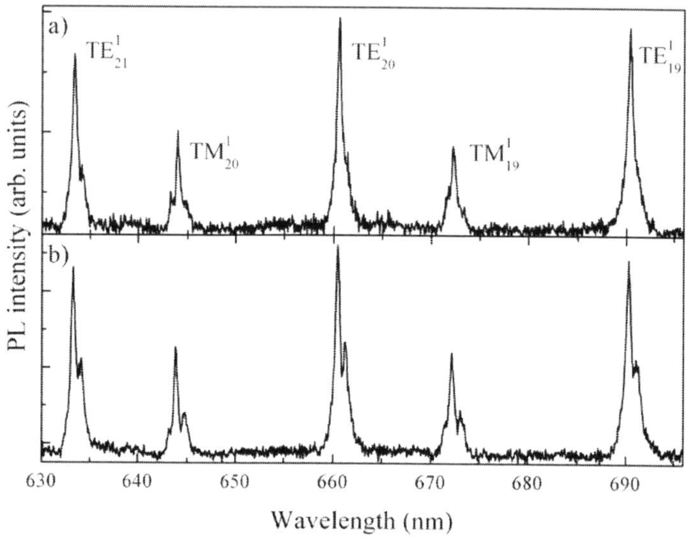


**Figure 5.10** PL spectra of microspheres with a thin PE-shell and one monolayer of CdTe NCs recorded while trapped in a cw laser beam under radiation pressure. The focus position is 1 μm above the microsphere (a) and near the substrate (b).

Figure 5.10 shows the results of two experiments. In these studies, the $Ar^+$ laser beam ($\lambda$ = 514.5 nm) simultaneously serves the purpose of an optical tweezers setup, confining the sphere and applying radiation pressure and as an excitation source for the optically pumped semiconductor NCs. A *xyz*-table with a spatial resolution of 0.1 μm/step allows an accurate focusing of the vertical laser beam through the high NA (0.9) microscope objective at 100× magnification. In the first case (Fig. 5.10a), the PL spectrum was recorded with a focus position of 1 μm above the microsphere. According to theoretical predictions (Fig. 5.9) it causes an upward radiation pressure to be exerted on the sphere. The output power of the $Ar^+$-ion laser is $P$ = 15 mW. The spectrum shows WGMs, which are typical for a sphere with a diameter of 3 μm. The stronger peaks in Fig. 5.10 were identified as TE modes, whereas the smaller peaks observed in the PL spectra are TM modes. All peaks are broadened and show some indication of weak splitting due to the radiation pressure and the weight of the sphere ($1.6 \times 10^{-12}$ N). When focused inside the sphere near the substrate as shown in Fig. 5.10b, a second peak arises slightly red-shifted with respect to the original mode. The broadening of the peak toward longer wavelength is in agreement with the theory, predicting that the deformation causes the lifting of the mode degeneracy and shifting the azimuthal modes to larger wavelength (Fig. 5.7).

A simple model based on Eq. (5.18) can verify that the observed peak structure is caused by lifting the mode degeneracy. Within this model, individual WGMs can be approximated by Lorentzian-shaped azimuthal peaks. The superposition of these modes is plotted in Fig. 5.11 including the original peak of the undeformed sphere.

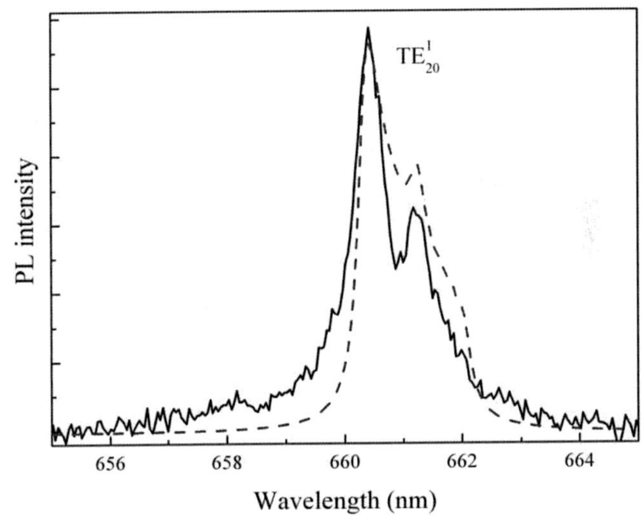


**Figure 5.11** Measured line shape of $TE_1^{20}$ mode (black line) and simulated peak shape (blue line).

This peak for the perfectly shaped sphere is added to the model due to an irregular deformation of the soft PE layer. As the solid sphere is not deformed but only the compressed PE layer, we assume that there is still a strong resonance at the original resonance frequency of the WGM, assigned to the modes that travel around the sphere circumference in the region that is not affected by the distortion. Several parameters for the modeled peak such as the linewidth of the resonances (estimated to be $\Delta\lambda$ = 0.3 nm) and the original resonance wavelength $\lambda_0$ are taken from the measured spectrum. The shape of the modeled peak is fitted to the measured peak by changing the ellipticity to the proper value. The result is shown in Fig. 5.11 demonstrating very good qualitative agreement between the calculated and measured line shapes of the peak. The estimated value of ellipticity $e = 8.5 \times 10^{-3}$ is equivalent to a decrease of the sphere radius by about 4 nm, which is a reasonable value taken into account the thickness of the PE shell.

## 5.4 Microcavity–Enhanced Raman Scattering and Anti-Stokes Emission from Monolayer of Nanocrystals

### 5.4.1 Cavity-Enhanced Raman Scattering

WGM microcavities are ideally suited for observing nonlinear optical effects with extremely low threshold powers. Enhancement of Raman scattering due to WGM resonances has been reported in number of experimental studies including microdroplets of ethanol and water mixed with dye molecules [32,33] and aerosols [34]. In these studies, the Raman scattering due to microcavity modes was termed cavity-enhanced Raman scattering (CERS). Recently Rakovich *et al.* reported on enhancement of Raman scattering from semiconductor quantum dots attached to a dielectric microsphere [35]. The major differences between microdroplets experiments and this study is that in liquid CERS the molecules of dye are inside the microdroplets filling their entire volume, whereas in [35] a shell of nanocrystals is placed just outside the microsphere, where the confined electromagnetic field associated with WGMs reaches its maximum.

The quality factor $Q$ of MF or polystyrene (PS) microspheres of 2 to 70 μm in diameter with a thin shell of CdTe NCs was found to be high enough to provide the feedback required for inelastic optical processes such as enhanced Raman scattering [35–37], which allows very efficient vibrational characterization of just a monolayer of semiconductor NCs coating the microsphere [35].

The measured Raman spectrum of a single MF/PE/CdTe microsphere of 2.2 μm size with a shell thickness of only one monolayer reveals pronounced features as shown in Fig. 5.12. The use of the Si substrate on which the microsphere is placed, provides the built-in standard of the Si TO mode at 520 cm$^{-1}$. The peak detected at 162.0 cm$^{-1}$ originates from the longitudinal optical (LO) phonon mode of the CdTe NCs monolayer. The frequency of this peak is smaller than the corresponding bulk value of 168 cm$^{-1}$ [38] by 6.0 cm$^{-1}$. This shift originates from two sources: a red shift due to confinement of optical phonons [39], and a blue shift caused by lattice contraction [40,41].

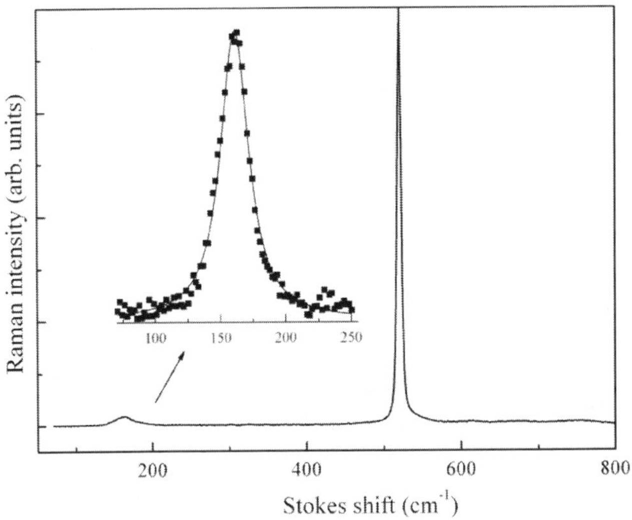


**Figure 5.12** Raman spectrum of a single MF/CdTe microsphere on a Si substrate. Excitation provided by an Ar$^+$-laser ($\lambda$ = 488 nm).

The 25.0 cm$^{-1}$ width of the LO phonon line for CdTe NCs in these experiments reflects the somewhat broad size distribution of the NCs. It is noteworthy that no Raman signal was observed at the same experimental conditions from a monolayer of CdTe NCs directly deposited on top of the Si wafer. This is compelling evidence that the Raman signal from a monolayer of CdTe NCs were observed due to strong coupling with the spherical microcavity.

As we discussed in Section 5.2, a strong increase in the value of the $Q$ factor is expected with an increasing radius of curvature of the microsphere. Increasing the size of microsphere leads to a strong reduction of mode volume and therefore to high light circulating intensities, due to the cavity buildup factor. In Fig. 5.13 the Raman spectrum of a microcavity-NCs system consisting of CdTe colloidal NCs coated onto a PS microsphere of 70 μm size is shown. The measured spectra at resonance excitation by an $Ar^{+}$-laser (488 nm) shows a number of peaks that are intrinsic to the PS (220, 620, 759, 793, 1001, and 1031 cm$^{-1}$) In addition to these lines, the cavity-induced enhancement of the Raman scattering allows for the observation of LO phonon mode from only a monolayer of CdTe NCs (166.0 cm$^{-1}$) (Fig. 5.13a, inset), red-shifted due to confinement of optical phonons [37].

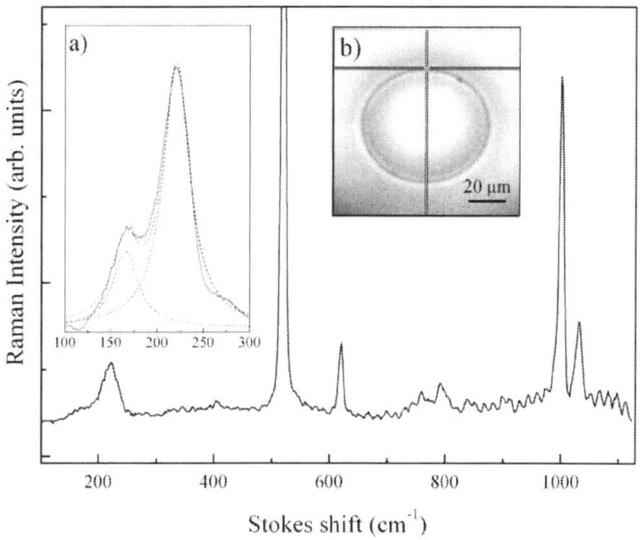


**Figure 5.13** Raman spectra of a single CdTe/PS microsphere on a Si substrate. Excitation by an $Ar^{+}$-laser ($\lambda$ = 488 nm). Inset: (a) The deconvolution of the line shape of the Raman spectra in the region of CdTe LO phonon mode using Lorentzian functions. (b) Microscope image of the CdTe/PS microsphere. The dark cross indicates the excitation position.

The most remarkable experimental observation is a periodic (ripple) structure with very narrow regular peaks in the Raman spectra of the PS/NCs microsphere, which can be clearly seen in Fig. 5.13. This structure corresponds to the WGM modes of the spherical microcavity and can best be detected by providing excitation at the rim of the microsphere (Fig. 5.13b, inset). The resonant internal field of a spherical cavity corresponding to high-$Q$ modes is mostly confined to the near-surface interior of the microsphere [42]. Therefore, a uniform intensity distribution within the rim of the microsphere in the volume determined by the WGM can be expected if the incident wave is resonant with a WGM. In such a situation edge illumination with a focused beam excites the WGM of a microsphere more uniformly and more efficiently than at excitation focused at the microsphere center [37,42].

Another feature, which can be clearly seen in Fig. 5.13 is that the intensity of WGM peaks decreases as they approach the excitation wavelength. The observed reduction of the peaks intensity in the Raman spectra of a PS/CdTe microsphere is due to absorption by NCs that are coupled to the relevant WGM. It is well known that absorption or gain or refractive index variation alter the $Q$ value of the spherical microcavity [43]. Because of the Stokes shift (30 nm) between the intrinsic PL peak and absorption of CdTe NCs, the absorption coefficient is reduced at the long-wavelength part of the PL band [44], allowing a higher $Q$ factor to be achieved in this spectral region.

Recently Ausman and Schatz offered an explanation of microcavity-enhanced Raman scattering in terms of electromagnetic "hot spots," the locally enhanced electric fields due to WGM resonances in silica microspheres [45]. Raman scattering enhancement factors of $10^3$–$10^4$ are found at positions slightly off the propagation axis when the incident excitation is matching a WGM resonance. Although experimental studies do not yield such increase in Raman intensity, apparently even modest enhancements can be used to detect monolayers of nanocrystal films [35,36] and subpicogram levels of organic materials [46] not readily detectable using conventional Raman spectroscopy.

### 5.4.2 Anti-Stokes Emission from a Single Spherical Microcavity with a Monolayer of CdTe Nanocrystals

Due to inhomogeneous broadening, the band edge emission from semiconductor nanocrystals was found to be strongly dependent on the excitation photon energy $E_{ph}$ when $E_{ph}$ is smaller than the fundamental absorption peak [47,48]. Resonant PL spectroscopy (size-selective excitation into the tail of the absorption onset) shows increasing fluorescence line narrowing with increasing excitation wavelength [49,50]. When $E_{ph}$ is close to PL band maximum, efficient thermally activated photon energy upconversion occurs leading to anti-Stokes photoluminescence (ASPL) [51–53]—the observation of an emission at energies higher than that of the excitation energy. The mechanism responsible for ASPL in semiconductor nanocrystals is briefly considered in Chapter 1. The excitation of ASPL can be extremely efficient, and recent advances in a number of applications including laser technology [54], three-dimensional microfabrication and optical storage [55,56], displays and optical limiting [54,57], imaging techniques [58], and biological caging [59], has renewed interest in upconversion luminescence [60]. Upconversion of semiconductor NCs combined with photon confinement in three-dimensional microcavities has the great potential to be useful in micro-laser technology, optical data storage, lighting, and bioimaging applications.

Figure 5.14b shows a room temperature spectrum of a single PS/CdTe microsphere on a Si substrate using low intensity non-resonance excitation by a HeNe laser. In that case, strong coupling between the WGM of the spherical microcavity and the electronic states of the CdTe NCs results in an enhanced luminescence contribution to the Raman signal simultaneously in both the Stokes and anti-Stokes spectral regions. The ASPL process is certainly highly efficient having intensity comparable to the Stokes PL as seen from Fig. 5.14b. It was found that the integrated intensity of ASPL has an almost linear dependence on the excitation intensity under weak or moderate excitation ($<200$ W/cm$^2$). This dependence is very similar to the behavior of ASPL in colloidal CdTe NCs where the progressive transition from Stokes PL into ASPL can be observed when changing the excitation wavelength to below the band-gap region [52,61–64].

A similar effect was recently reported in small (2 μm) microspheres with a thin shell of semiconductor nanocrystals, and explained by multiphonon-assisted excitation of an electron from the ground state to the excited state through the mediation of the shallow trap levels [35]. In the case of colloidal NCs, such a low cross section mechanism, like anti-Stokes excitation, can be only efficient in samples with high quantum yields [64,65]. The observation of ASPL from a PS/CdTe microsphere can be attributed to the optical feedback *via* the microcavity with a WGM structure, which leads to an increased probability of energy transfer to the emitting species.

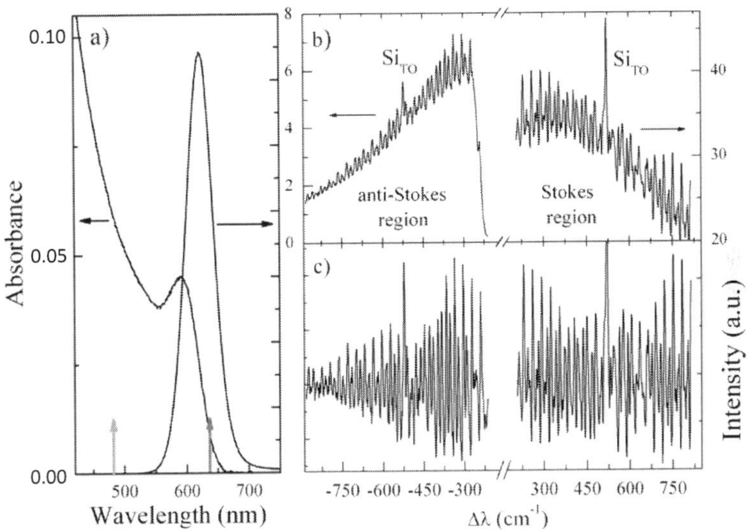


**Figure 5.14** Room temperature absorption and PL spectra of CdTe NCs in water (a). Arrows indicate the excitation wavelength used in micro-PL and Raman experiments. Raman spectrum of a single PS/CdTe microsphere on a Si substrate. Excitation by HeNe laser ($\lambda$ = 632.8 nm) before (b) and after (c) PL background subtraction.

As a result of the very high PL quantum efficiency of the NCs, the WGM peaks in the spectra presented in Fig. 5.14b are superimposed on a broad background PL signal. In order to show more clearly the WGM structure itself, this background may be subtracted using a multi-Gaussian function (Fig. 5.14c). This procedure allowed us to reveal one more feature, which is typical for spherical microcavities, a slowly oscillating structure caused by interference between light diffracted and transmitted at the poles of microsphere [66]. As shown in Fig. 5.14c, the WGM peaks are in fact grouped together and a periodic interference structure can be clearly seen both in the Stokes and anti-Stokes spectral regions with a period of ~340 $cm^{-1}$ and ~230 $cm^{-1}$, respectively.

Similar to the PL spectra of a single microcavity, the Raman spectra (Fig. 5.14b,c) show a sequence of sharp peaks, which only occur at discrete frequencies depending on the refractive index, $n_r$, and the radius $R$ of the microsphere. The modes in the spectrum in Fig. 5.14b are arranged in pairs of two pronounced peaks where the TE mode corresponds to the peak with higher intensity and the TM mode to the smaller peak, which follows from polarization experiments. However, in contrast to the case of small (<10 μm size) microcavities accurate identification of the WGMs in 70 μm microspheres is impossible due to strong overlap of resonant peaks and high $Q$ factor.

Figure 5.15b shows an enlargement of the measured spectrum (Fig. 5.14b) where the periodic structure of WGM peaks can be seen in more detail, demonstrating that the modes in the PL spectrum are arranged in pairs of two pronounced peaks one of higher intensity and a second smaller peak. Moreover, a few extra very small peaks can be distinguished in the spectral region between them. A Fourier analysis of the spectra allows us to gain more insight into the WGM structure in the microcavity making it possible to investigate the periodicity more thoroughly.

In the spectral frequency interval 0–3.5 $nm^{-1}$, we observed strong peaks corresponding to a periodicity of 1.31 nm, 0.65 nm, 0.44 nm, and 0.33 nm (see bars in Fig. 5.15a,c). The highest periodicity value could be assigned to the free spectral range (FSR) between modes of the same polarization with radial order numbers $n = 1$. Because transverse electric modes (TE) have normally a higher quality factor than transverse magnetic (TM) modes, we can attribute the stronger peaks in Fig. 5.15b to the TE modes of the WGM. The good agreement between the measured FSR of 1.32 nm between the TE modes supports this hypothesis. In turn, the periodicity of 0.65 nm is attributed to the FSR between modes of different polarizations (i.e., between adjacent TE and TM modes) again in agreement with measured modes separation (Fig. 5.15b). Periodicities of 0.44 and 0.33 nm, obtained from the Fourier analysis, are indicative of the TE and TM modes with radial order numbers, which are greater than one.

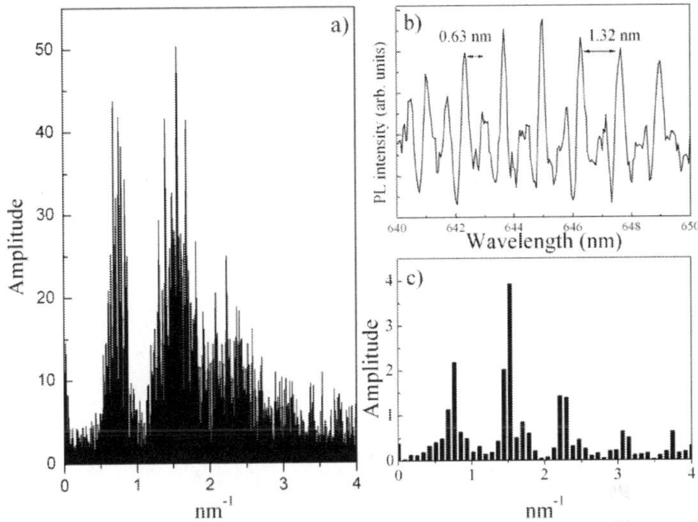


**Figure 5.15** Result of fast Fourier analysis over the whole spectral region (a). Expansion of the recorded spectrum (b) and corresponding result of fast Fourier analysis (c). Arrows indicate the free spectral range (FSR) and TE/TM mode splitting.

## 5.5 Mode Modification in Spectra of Coupled Spherical Microcavities with Semiconductor Nanocrystals

As discussed in Section 5.1, the resonant internal field of a spherical cavity is not completely confined to the interior of the microsphere. Depending on the size of the microsphere, the evanescent field can extend into the surroundings up to a couple of micrometers. It was recently recognized that the partial delocalization of Mie resonance states is of great importance because it suggests a possibility for coherent coupling between WGMs of two adjacent spherical particles with closely matched sizes. Such a system of coherently coupled photonic atoms may be called a "photonic molecule" (PM) [1,67] and can be employed in order to manipulate photons in the micrometer length scale. In analogy to the formation of molecular electronic orbits, the tight binding approximation provides two combinations for the electromagnetic field in a system of interacting microspheres: bonding (BN) and antibonding states (Fig. 5.16) [67–70].

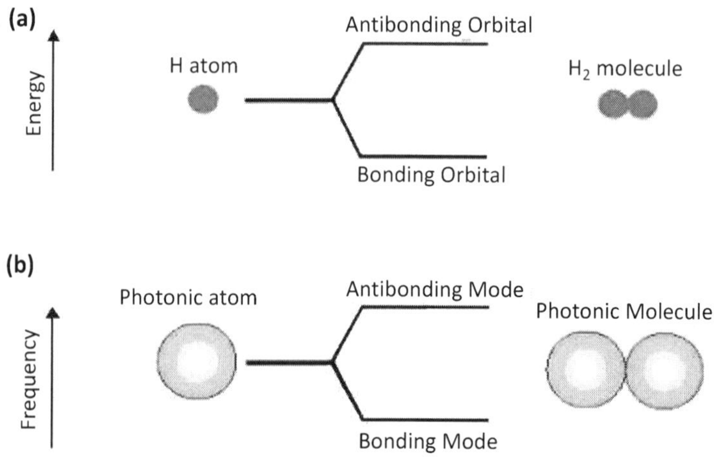


**Figure 5.16** Schematic of the splitting of the energy levels of hydrogen atoms (a). By analogy, the WGMs with the same resonant frequency split into two coupled modes of the photonic molecule (b).

Experimentally, the coupling of the photon modes of individual microspheres in the PM can cause narrow resonance of a photonic atom to split into two modes of lower $Q$ factor [68]. This phenomenon has been demonstrated in a system of two square, photonic dots coupled by a narrow waveguide channel [67], in a dye-stained bisphere system [70–72], in photonic dots doped by CdSe [73] and CdTe [74–76] NCs and in chains of polymer-blend microparticles [77]. However recent theoretical considerations [69] and experimental studies [74,75] reveal a complex internal distribution of the density of photonic states of the PM originating from lifting of the degeneracy of PM modes with respect to the azimuthal index.

### 5.5.1 Confined Photonic States in Weakly Coupled WGM Microcavities with CdTe Nanocrystals

The simplest experimental approach to investigate the propagation modes in interacting spherical microcavities can be realized by scanning a sample along the longitudinal axis of the PM using an excitation/detection geometry as presented in Fig. 5.17 [75].

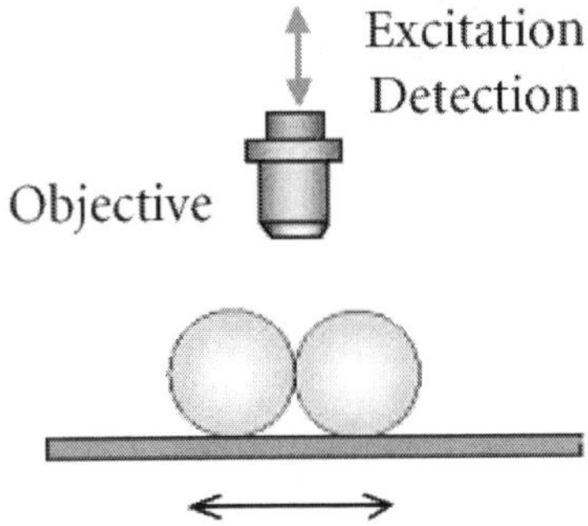


**Figure 5.17** Experimental geometry for scanning a PM along the longitudinal axis.

Figure 5.18 shows experimentally measured PL spectra of coupled MF/CdTe microspheres and the results of WGM identification using the algorithm described in Section 5.2. The shift in position of the $TE^1_{22}$, $TE^1_{23}$, $TE^1_{21}$, and $TE^1_{22}$ modes of single microspheres, which can be clearly seen in Fig. 5.18 is the result of a difference in the size of the two microspheres of only 11 nm as follows from calculations based on Mie theory.

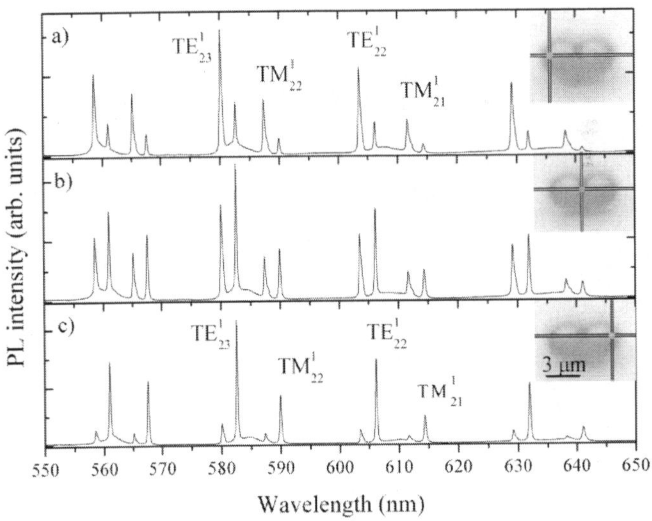


**Figure 5.18** PL spectra of a PM with excitation and detection at three different positions along its longitudinal axis. Insets: microscope images of the PM, with the cross-hairs indicating the excitation-detection position.

When the excitation is provided at the contact point between microspheres (Fig. 5.18b), the excited and emitted light propagate efficiently along the longitudinal axis of the PM and the contribution of both microcavities can be clearly seen in the PL spectra at discrete WGM wavelengths. This contribution of PL signal from individual microcavities can be switched by changing the excitation position from the left to the right rim of the PM (Fig. 5.18a,c). The above experiments involved an excitation beam with incidence perpendicular to the longitudinal axis of the PM. For a given excitation and detection configuration no other new peaks were observed in the PL spectra of the PM compared with that of single spheres, which is indication of weak intermode coupling. The lack of intermode coupling in this configuration (Fig. 5.18) is not surprising because the coupling between electromagnetic fields of the spheres is expected to be minimum when the excitation and detection are set perpendicular to the axis of the PM [70]. However, the observed redistribution of intensity between the components of WGM double structure clearly demonstrates the waveguiding of the light along the PM and the possibility of wavelength switching in the PM depending on the excitation-detection geometry.

### 5.5.2 Confined Optical Modes in Small Photonic Molecules in Strong Coupling Regime

Theoretical considerations [68–70] show that the intersphere coupling is expected to be a maximum for the pair of modes whose orbitals include the contact point between microspheres and lie in the same plane. Taking this into account, the PL intensity of the coupled modes is anticipated to be a maximum in the direction parallel to the PM axis and the signal from the coupled intersphere modes should be more pronounced in the parallel configuration than in the perpendicular one.

Recently, a new approach to control the alignment of the spheres constituting the PM have been developed [74,75] utilizing a polystyrene substrate containing a three-dimensionally ordered array of pores of ~5 μm in size prepared through a thermocapillary convection [78]. The ordered structures are formed by evaporating solutions of polystyrene in a volatile solvent, in the presence of moisture with forced airflow across the solution surface. A hexagonally packed array of holes (microwells) of 3–5 μm depth then forms on the surface of the polymer (Fig. 5.19a). Only one pair of the 3 μm microspheres can be accommodated within each microwell, and the axis of the PM is close to the surface normal (Fig. 5.19b).

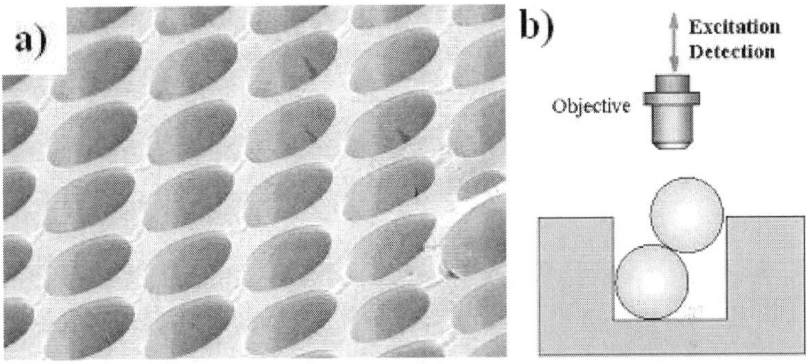


**Figure 5.19** Polystyrene film with an hexagonally ordered array of air holes with diameters of about 5 μm (a), and experimental geometry for off-axis or parallel excitation and detection (b).

Figure 5.20 shows the PL spectrum of a PM accommodated in a microwell and the spectra of the individual microspheres prior to being manipulated into the microwell. The pronounced doublet structure in the spectrum presented in Fig. 5.20a, coinciding with the spectral positions of the WGM of individual microspheres (Fig. 5.20b), is a result of the overlap of uncoupled modes of the two spheres. The shift in position of the WGM, which can be clearly seen in Fig. 5.20a,b, is a result of a difference in the size of the two microspheres of only 25 nm. However, in contrast to the spectrum in Fig. 5.20a, this doublet is accompanied by two extra relatively broad peaks (indicated by arrows in Fig. 5.20b, which are indicative of strong mode coupling in the PM). The appearance of these two satellites can be interpreted as a result of the formation of BN and ABN orbitals in the PM [67] with the ABN peak observed at lower wavelength than the BN one. In terms of cooperative scattering theory [68], the observed satellites originate from the removal of the WGM degeneracy with respect to the *m* index. The line shape of the satellite lines reflects the energy distribution among the coupled modes, because modes with different combinations of *m* can contribute to the PL spectra. The observation of a broader ABN peak, relative to the BN peak, reflects the decrease of the quality factor of the PM when compared with that of a single sphere presumably due to the interaction with more dissipative modes of lower *l* [70]. The deconvolution of the line shape of resonances belonging to $TE^1_{22}$ WGM (Fig. 5.20a,b) using Lorentzian functions show that the quality factors of PM peaks are ~8 times smaller than *Q* factors of noninteracting microspheres.

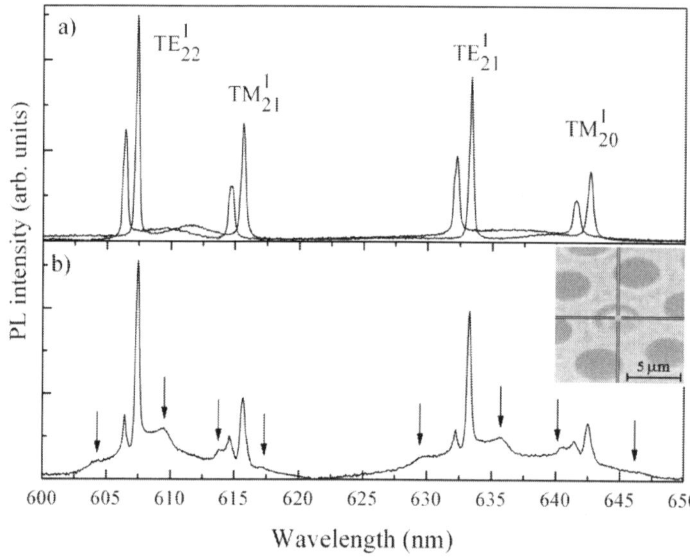


**Figure 5.20** PL spectrum of non-interacting microspheres (a) and PL spectrum of PM accommodated in a microwell (b). Arrows indicate the coupled modes. Inset: The dark cross indicates the excitation position.

## 5.5.3 Optical Modes in Photonic Molecules Formed from Spherical WGM Microcavities

Photonic states in the PM can be described as a linear combination of the Mie resonance states of each microsphere [68] or using a more general phenomenological tight-binding model [69], which has also been used recently to reproduce the band structure of two-dimensional photonic band-gap lattices composed of cylinders [5].

For a given WGM polarization, the value of the splitting between the BN and ABN modes of the PM can be obtained using the maximum term approximation of the single-mode tight-binding (SMTB) method [69]

$$\Delta x_m = 2\Gamma_n \left|A_{n,m}(x_n)\right|, \tag{5.25}$$

where $x_n = 2\pi R/\lambda_n$ is the size parameter of a resonance with mode number $n$, $\lambda_n$ is the corresponding resonant wavelength, $\Gamma_n$ is the width of mode $n$ in a single sphere, which can be calculated with Mie-theory. For a given radius of a single sphere $R$, the coefficient $A_{n,m}$ can be calculated as follows:

$$A_{n,m}(x_n) = -2n(-1)^{n+m} h_{2n}^{(1)}(k_0 Z) \times \sqrt{\frac{2n}{\pi(n+m)(n-m)} \times \frac{n^{2n}}{(n+m)^{n+m}(n-m)^{n-m}}}, \tag{5.26}$$

where $k_0 = x_n/R$ is the wavenumber, $Z$ is the characteristic length [69] and the spherical Henkel function of the first can be estimated from

$$h_{2n}^{(1)}(k_0 Z) \cong -i\frac{\exp\left[(2n+1/2)(\alpha - \tanh\alpha)\right]}{(n+1/2)\sqrt{\operatorname{sech}\alpha \tanh\alpha}}; \tag{5.27}$$

Here, $\alpha$ is defined by $k_0 Z = (2n + 1/2)/\cosh\alpha$. In the simplest approach the positions of the $m$-resonances can be estimated as $x_m = x \pm \Delta x_m/2$.

Due to the high PL efficiency of the CdTe NCs and the coupling of electronic transitions of NCs to the PM resonances, the BN and ABN branches were detected in a wide spectral region from 525 to 725 nm, corresponding to the full spectral range of the NC emission. This allows the possibility of estimating the magnitude of spectral spacing between bonding and antibonding branches (Fig. 5.20) as a function of angular mode number for comparison with theoretical calculations (Fig. 5.21).

Applying Eqs. (5.25)–(5.27), we can see that the modeled data presented in Fig. 5.21 display the behavior that is in good agreement with experimental data. A continuous decrease in splitting between BN and ABN branches value with increasing $n$ can be seen in Fig. 5.21, with TE modes splitting being always higher than that of TM modes. Remarkably, the theory predicts the same behavior. Although the maximum term approximation of SMTB method overestimates the splitting (Fig. 5.21), it is clear that this approach provides a good qualitative guide for the analysis of a variety of phenomena observed in the PM structure.

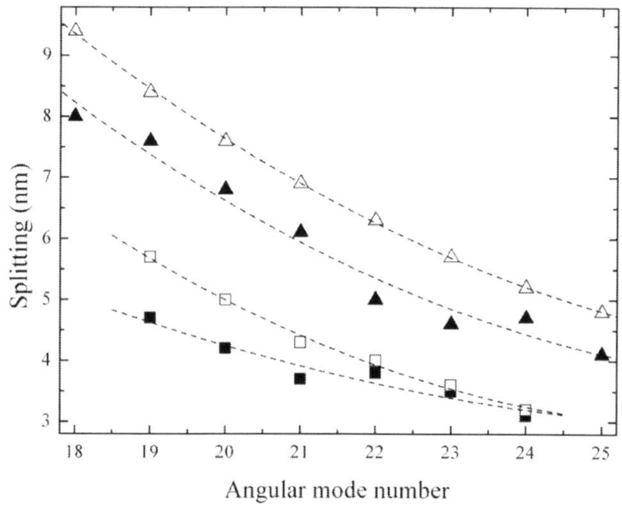


**Figure 5.21** Magnitude of spectral spacing between bonding and anti-bonding branches estimated from micro-PL spectra of a PM versus the angular mode number for TE (solid up-triangles) and TM (solid squares) modes. Open symbols show the result of theoretical calculations based on the maximum term approximation of the single-mode tight-binding method (5.25–5.27). The dashed lines are a guide to the eye.

It is worth noting that observed disagreement between the calculated and experimental values for the PM mode splitting (Fig. 5.21) may also have its origin in the fact that the magnitudes of splitting were estimated as a spectral distance between the maxima of ABN and BN peaks (indicated by arrows in Fig. 5.20b). However, detailed consideration of coherent mode coupling in PMs using the SMTB approach shows that in fact the BN and ABN branches consist of a number of very narrow peaks, which are due to the presence of $m \neq \pm 1$ components [69]. The total number of these sharp peaks, originating from a certain mode $n$, is governed by the actual degeneracy of the Mie resonances, which in the approach of the normal mode concept is $n + 1$. (Due to the dependence of the mode coupling on the orbital plane orientation, the interaction is limited between modes of only the same $m$, no degeneracy is removed between $m$ and $-m$ and the new degeneracy of PM modes is now $n + 1$ with $m$ running from 0 to $n$.) In calculations within the SMTB model, the splittings were estimated as a spectral distance between outer peaks assigned to modes with $m = \pm 1$, which do not correspond to the maximum of envelope of the PM modes and therefore this may cause a discrepancy between the estimated and observed values of the mode splitting. The fine structure of BN and ABN modes of PM is of great interest to the experimentalists because it suggests a manifold of applications particularly in the field of information processing [74,79]. However, in order to observe this phenomenon two conditions should be met. First, the spacing between $m$-resonances forming the BN and ABN modes of the PM strongly depends on the angle of incidence of the electromagnetic wave $\theta$ [69,70,74]. For $\theta = 0°$, when the incident light propagates parallel to the longitudinal axis of the PM, the incident wave can preferentially excite $m = \pm 1$ modes. The interaction with other dissipative modes of lower $l$ causes the broadening of BN and ABN peaks and decreases the $Q$ factor relative to that of single spheres, as was observed in a number of papers [68,70,74,75]. In the perpendicular configuration, when $\theta = 90°$, the intermode coupling and spacing between $m$-resonances are expected to be minimal. In other words, in the case of perpendicular orientation of the PM with respect to excitation, all $m$-modes would merge into broad BN and ABN features and their fine structure would not be easily recognized. An intermediate situation can be reached only for off-axis excitation with $0° < \theta < 90°$. Second, in order to reveal the $m \neq \pm 1$ components the interacting cavities should have not only similar size, but also similar $Q$ factors. If the resonances of the two cavities are of greatly different width, the coherent coupling is disturbed (at least to some extent) and therefore the fine structure of PM modes cannot be detected.

One can see from the inset in Fig. 5.20b that excitation and detection in this case was provided at the center of the upper sphere forming the PM i.e., along the PM axis. In order to reveal the $m \neq \pm 1$ components in the PL spectra of the MF/CdTe PM the WGM structure should be investigated at different excitation (and detection) positions across the upper sphere.

Figure 5.22a shows PL spectra of a PM formed by two almost identical microspheres with sizes of 3.0168 μm and 3.0189 μm measured for excitation (and signal collection) at the right edge of the upper microsphere (inset, Fig. 5.22a). The presented PL spectra clearly reveal major features unique to strong coherent coupling between the photonic states of the two microspheres forming the PM. One can clearly see a number of narrow peaks grouping on both sides arising from TE and TM resonances of the individual spheres and forming BN and ABN modes of the PM. The origin of this fine structure of the BN and ABN modes lies in the lifting of the mode degeneracy in the PM [69,74] and therefore this multi peak structure would not be radically altered in scattering spectra or even in lasing, or stimulated Raman scattering in the same way that in a semiconductor laser the lasing does not alter the cavity resonances.

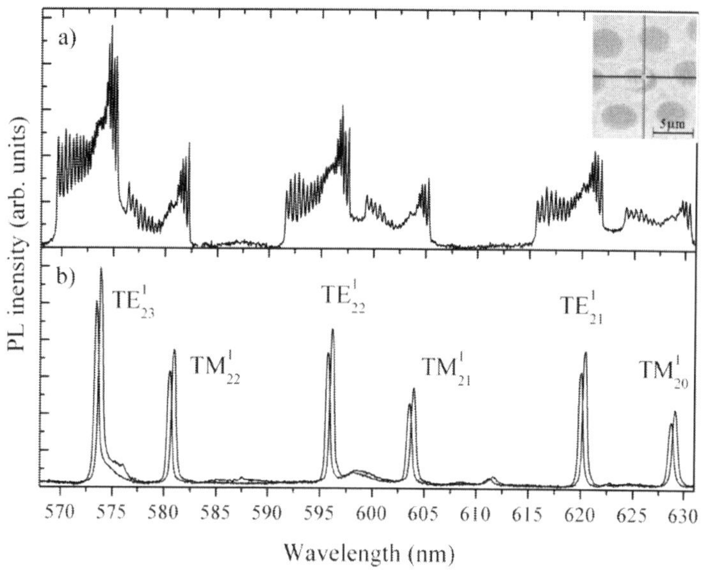


**Figure 5.22** PL spectra of the PM formed by two almost identical microspheres with off-axis excitation and detection (a), and PL spectra of non-interacting microspheres (b). Inset: microscope image of the PM in the microwell. The dark cross indicates the excitation position.

It is noteworthy that the number of experimentally resolved peaks increases with the angular mode number both for TE and TM modes, and is very close to the *n* value, although never in excess of *n*. Indeed we have observed 19 peaks in the spectral region of $TE^1_{20}$ (19 peaks for $TE^1_{20}$), 21 peaks around the $TE^1_{21}$ resonance (20 peaks for $TE^1_{21}$), 21 peaks in the region of the $TE^1_{22}$ resonance (21 peaks for $TE^1_{22}$), 22 peaks for $TE^1_{23}$ (22 peaks for $TE^1_{23}$) and 23 peaks around the $TE^1{}_{24}$ resonance. Note that the spectral region occupied by a set of these narrow resonances is much wider than linewidth of WGMs of non-interacting microspheres (Fig. 5.22b).

The deconvolution of the line shape of the *m*-resonances of the PM using Lorentzian functions shows that *m*-resonances of the BN branch are always sharper than that of ABN one, providing a higher quality factor *Q* value for these modes and therefore a higher photon lifetime in the resonant modes $\tau$. However, the most remarkable experimental fact is that the *Q* factor of the *m*-resonances in the spectra of the PM exceeds the *Q* value of single noninteracting microspheres, suggesting the respective modification of photon lifetime in the PM ($\tau_{PM}$) relative to the photon storage time in a single spherical microcavity before contact ($\tau_{SS}$). These two facts along with estimated value of BN/ABN splitting (~5–7 nm) implies the possibility for development of a new PM-based photonic device such as an optical delay line with controllable spectral and temporal tenability [79].

Figure 5.23 shows the spectral distribution of the ratio between photon lifetime of *m*-modes of PM and photon storage time in WGMs of single sphere calculated from the corresponding values of mode linewidth in the region of $TE^1_{21}$ and $TE^1_{20}$ resonances demonstrating a threefold photon storage enhancement in the PM. It is evident from Fig. 5.23 that the interaction between spherical microcavities results in periodic group delay spectra with peaks occurring at each of the *m*-resonant frequencies with longer delay times for higher *m*-values, which implies that the spectral components near these *m*-resonance spend more time traveling within the PM.

The relative increase in photon storage time $R = \tau_{PM}/\tau_{SS}$ shows an almost linear dependence on the azimuthal number *m* for given mode number *n* (Fig. 5.24), however it turned out that slope of this dependence ($\Delta R/\Delta m$) in turn varies with *n* (Fig. 5.24, inset).

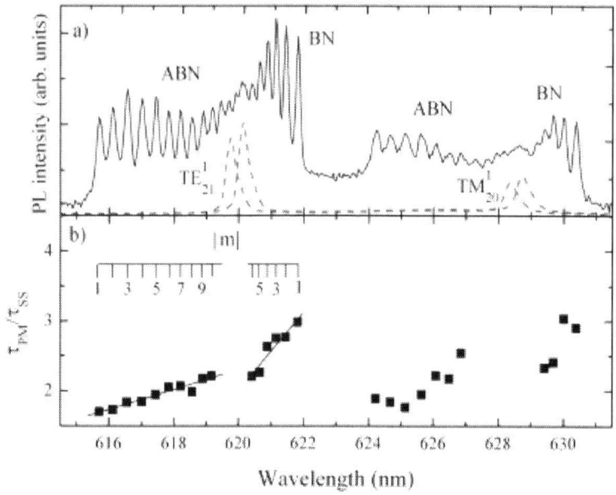


**Figure 5.23** (a) PL spectra of the PM (solid line) formed by two almost identical melamine-formaldehyde microspheres with off-axis excitation and detection. Dashed lines show the PL spectra of the non-interacting microspheres. (b) Ratio between the photon lifetime of the $m$-modes of the photonic molecule and that of single spheres. The solid lines are the result of a linear fit.

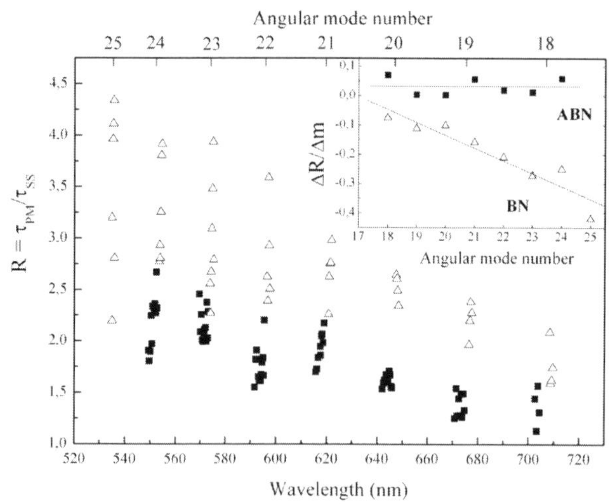


**Figure 5.24** Ratio between the photon lifetime of $m$-modes of the photonic molecule ($\tau_{\mathrm{PM}}$) and the photon storage time in the WGM of the single sphere ($\tau_{\mathrm{SS}}$). The inset shows the $n$-dependence of the $\Delta R/\Delta m$ slopes for BN and ABN modes.

In all the experiments described here, the strong coupling phenomenon in the spectra of the PM was investigated in the visible and near-infrared spectral regions. However recent achievements in the development of infra-red-emitting colloidal nanocrystals can meet the first basic requirement for PM devises operating at such wavelengths, i.e., in the basic communication windows around 1300 nm and 1550 nm [80,81]. Figure 5.25a shows the values of spacing between adjacent *m*-resonances calculated in spectral region of the C- and L-bands for PMs formed by two identical $SiO_2$ microspheres with sizes of 10 μm (3), 12 μm (2) and 16 μm (1). The standard for channel spacing in today's optical communication systems 40–100 GHz was taken into account deciding on microsphere sizes. In this model each *m*-resonance in the PM signal can be considered a channel with spacing between channels depending on the size of the microspheres and the azimuthal mode number.

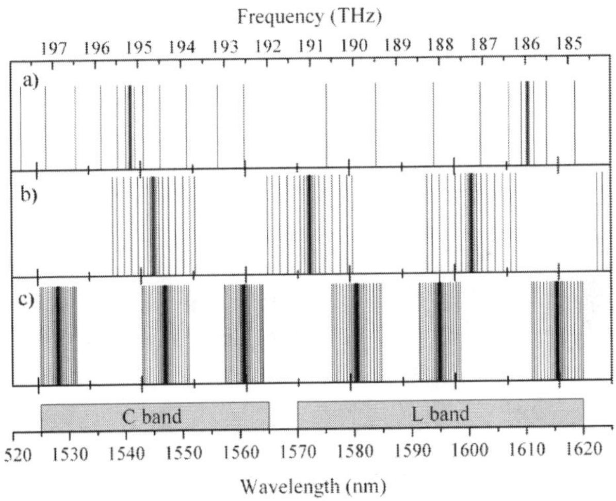


**Figure 5.25** Spectra covering two major communication bands calculated for a PM formed by two identical $SiO_2$ microspheres with a refractive index $n_r = 1.4$ and sizes of 5 μm (a), 10 μm (b), and 16 μm (c).

For example for 10 μm individual microcavities, only one TE mode ($\lambda_{24}$ = 1542.4 nm) is found in the C-band (Fig. 5.25b). The maximum splitting between BN and ABN modes (i.e., for $m$ = (1) of 15.7 nm was estimated in this case. Calculating the positions of resonances for a single 16 μm microsphere one can find two WGMs in the region of the C-band: $TE_{40}$ ($\lambda$ = 1544.6 nm) and $TM_{39}$ ($\lambda$ = 1559.6 nm). Maximum splitting in this case is much smaller: 8.8 nm for the $TE_{40}$ WGM but accommodates a larger number of *m*-resonances. These results clearly demonstrate that two coupled WGM spherical microcavities can generate modal structure with controllable number of peaks distributed across the major communication band. In contrast to the case of single spheres, the bandpass of the PM device can be as wide as the width of all communication bands and can be controlled by the size of spherical microcavities forming PM.

Figure 5.26b shows the estimated distribution of photon storage times between *m*-modes of the PM taking values of the photon storage time in single microspheres to be $\tau_{SS}$ = 55 ps for $TE^1_{40}$, 3 ps for $TE^1_{29}$ and 0.9 ps for $TE^1_{24}$ as determined from calculations based on the Mie-theory. It can be anticipated that the values of photon lifetime, $\tau_{PM}$, calculated in such a way are underestimated, because of the difference in *Q* factor achievable for 3 μm spheres and 10–16 μm ones. Recent experimental studies clearly show a strong dependence of the *Q* factors of *m*-resonances (and therefore photon storage times) on quality factor of individual microspheres [74,75].

The observed non-uniformity of the spacing between *m*-resonances (Figs. 5.24–5.26) provides a unique possibility to control the distribution of photon storage times in a particular spectral region (Fig. 5.26b). For all sizes of interacting microspheres, an initial increase of the spacing within the region of first three *m*-resonances was observed followed by a dip and then an increase in spacing again (Fig. 5.26a). This theoretical result is in good qualitative agreement with experimental data [74]. For the smallest PM size, 18 *m*-modes are available within a window of 12–200 GHz. The smaller *Q* factor of the WGMs in the individual spheres forming the PM results in a relatively moderate increase in photon lifetime—from 1.5 ps (obtained for ABN resonance with *m* = (1) up to 7 ps calculated for the corresponding *m*-resonance of BN mode of PM. For microspheres of larger size, the number of available modes increases. Indeed, for 12 μm coupled spheres, 20 *m*-resonances fit into above indicated spacing window, with the biggest value of intermode spacing ~130 GHz. Photon storage times distributed between these *m*-modes increase from 5 ps up to 32 ps. The upper limit of intermode spacing drops even more for PMs formed from 16 μm microspheres. In that case, 22 $m$-resonances with spacing between 12 and 76 GHz can be seen in Fig. 5.26a, providing discrete storage times found to be distributed between 94 ps and 960 ps. Remarkably, this maximum photon storage time provided by two interacting microspheres of 16 μm size corresponds to a ~15 cm length of a silica-glass waveguide delay line.

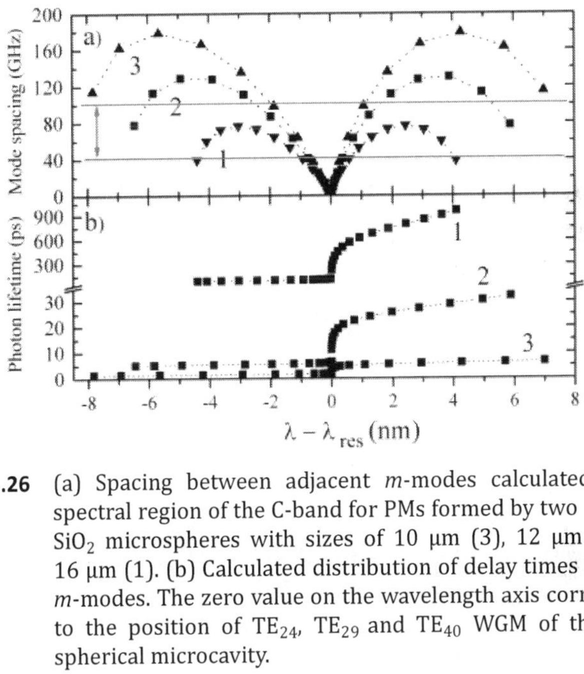


**Figure 5.26** (a) Spacing between adjacent $m$-modes calculated in the spectral region of the C-band for PMs formed by two identical $SiO_2$ microspheres with sizes of 10 μm (3), 12 μm (2) and 16 μm (1). (b) Calculated distribution of delay times between $m$-modes. The zero value on the wavelength axis corresponds to the position of $TE_{24}$, $TE_{29}$ and $TE_{40}$ WGM of the single spherical microcavity.

Based on theoretical consideration of the internal field in the PM [69] and a number of experimental works [70,73,74], it is anticipated that the size of interacting microcavities is probably not the only parameter that allows us to control the spacing between split modes and distribution of photon storage times. First, the efficiency of coupling between spherical microcavities forming the PM strongly depends on the spacing between microspheres: The splitting of the BN and ABN modes decreases exponentially with the increasing separation [69,79] For the present values of microsphere sizes, mode numbers and small $d$ the condition

$2n + 1 > x_n(2 + d/a)$ is satisfied and the SMTB method yields an approximate equation for coefficient $A_{n,m}(x_n,d)$:

$$A_{n,m}(x_n, d) = A_{n,m}(x_n) \times \exp\left[-x_n \frac{d}{a}\sqrt{\left(\frac{4n+1}{4x_n}\right)^2 - 1}\right]. \tag{5.28}$$

Replacing $A_{n,m}$ in Eqs. (5.25) by $A_{n,m}(x_n,d)$, we can estimate mode spacing as a function of $d$.

Taking, for example, the PM formed from 10 μm microspheres ($TE_{24}$ single-sphere resonance within C-band) the maximum splitting between BN and ABN modes shows a decrease by a factor 1.3 as the spacing between spheres increases from 0 to 0.1 μm. (Fig. 5.27) Corresponding modification of the $A_{n,m}(x_n,d)$ coefficient results in a significant decrease of the values of channel spacing. While the delay time and number of channels available between 12–200 GHz are unaffected, the upper limit of channel spacings drops down to 139 GHz (22% decrease). For two 10 μm spheres separated by $d$ = 0.23 μm spacings of all 18 channels are distributed between 12 and 100 GHz.

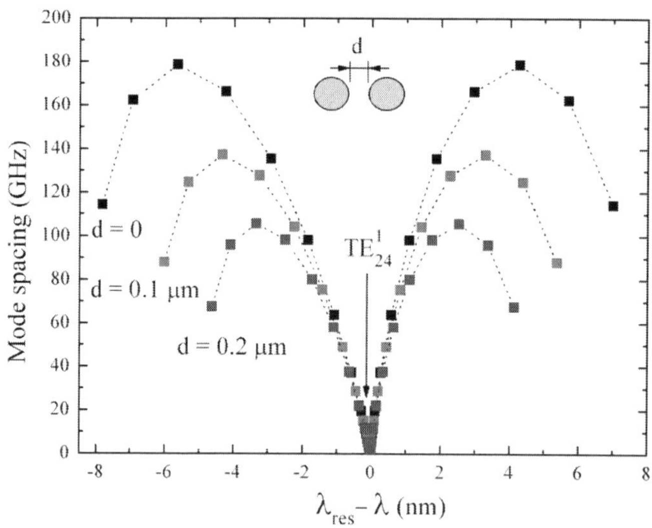


**Figure 5.27** Spacing between adjacent $m$-modes calculated for a PM formed by two identical $SiO_2$ microspheres with a size of 10 μm separated by a distance $d$.

The next possibility to control the spacing between channels emerges from the strong dependence of split parameters of PM modes on the angle of incidence of electromagnetic wave $\theta$. For $\theta = 0°$, when the incident light propagates parallel to the longitudinal axis of the PM, the incident wave can only excite $m = \pm 1$ modes [69,70,74]. Interaction with others dissipative modes of lower $n$ causes the broadening of BN And ABN the peaks and their $Q$ factor is lower than that of single spheres, as was observed in number of studies [68,70,74,75]. In the perpendicular configuration, when $\theta = 90°$ intermode coupling is expected to be minimal and the fine structure of PM modes cannot be easily resolved. In that case however, the $Q$ factor of BN and ABN modes can be at least comparable to that of non-interacting microspheres, presumably due to disappearance of one wing of these PM branches. The bandpass spectrum was demonstrated to be much more complicated for off-axis excitation with $0° < \theta < 90°$ showing a number of small peaks between upper lower peaks of $m = \pm 1$ [70,74,75]. Although further investigations need to be done on the angular dependence of the fine structure of the PM modes, recent investigations point clearly to the last configuration as the basis for the development of a PM-based optical delay device with controllable spectral and temporal tunability.

Finally, it remains to be noted that the parameters of the intercavity coupling in the PM can be governed also by detuning of the microsphere sizes. Forming PM from pairs of microspheres of slightly different sizes one can manipulate the spectral position of BN and ABN modes controlling the symmetry of the splitting.

## 5.6 Summary and Outlook

The universality and simplicity of the LBL technique enabled the preparation of composite photonic systems consisting of a spherical microcavity and semiconductor nanocrystals of CdTe. This system is highly efficient in coupling out light from the NCs making feasible a new method for photonic mode coupling in photonic molecules. The study of microcavity/NCs structures demonstrated here, reveals many fascinating questions and fundamental problems that appear when merging two major concepts of solid-state physics, complete 3D electronic and photonic confinement, in one structure. The dependence of WGM parameters on the characteristics of both interacting and noninteracting microcavities suggests a manifold of photonic applications, in particular in the fields of bio-sensing and quantum information processing.

## Acknowledgments

We thank Prof. J. J. Boland, Prof. A. Eychmüller, Prof. Y. Gun'ko, and Dr. N. Gaponik for helpful discussions and help with the experimental arrangements. This work was supported by Science Foundation Ireland (SFI) under its Project 08/IN.1/I1862.

## References

1. Arnold, S. (2001). Microspheres, photonic atoms and the physics of nothing, *Am. Sci.*, **89**, pp. 414–421.
2. Rayleigh (1912). *Scientific Papers* ed. (Cambridge University, Cambridge, England) pp. 617–620.
3. Chang, R. K., and Campillo, A. J. (1996). *Optical Processes in Microcavities*. Singapore, World Scientific.
4. Eisberg, P., and Resnick, R. (1985). *Quantum Physics of Atoms, Molecules, Solids, Nuclei and Particles*. New York, John Wiley & Sons, Inc.
5. Lidorikis, E., Sigalas, M. M., Economou, E. N., and Soukoulis, C. M. (1998). Tight-binding parametrization for photonic band gap materials, *Phys. Rev. Lett.*, **81**, pp. 1405–1408.
6. Bohren, C. F., and Huffman, D. R. (1983). *Absorption and Scattering of Light by Small Particles*. New York; Chichester, Wiley.
7. Chylek, P. (1990). Resonance structure of Mie scattering: distance between resonances., *J. Opt. Soc. Am. A*, **7**, pp. 1609–1613.
8. Kerker, M. (1969). *The Scattering of Light and Other Electromagnetic Radiation*. New York; London Academic P.
9. Van de Hulst, H. C. (1957). *Light Scattering by Small Particles*, New York, John Wiley & Sons.
10. Chylek, P. (1976). Partial-wave resonances and the ripple structure in the Mie normalized extinction cross section, *J. Opt. Soc. Am.*, **66**, pp. 285–287.
11. Rakovich, Y. P., and Donegan, J. F. (2010). Photonic atoms and molecules, *Laser Photon. Rev.*, **4**, pp. 179–191.

12. Decher, G., and Schlenoff, J. B. (2002). *Multilayer Thin Films: Sequential Assembly of Nanocomposite Materials. Weinheim*, Wiley-VCH.
13. Chew, H. (1987). Transition rates of atoms near spherical surfaces, *J. Chem. Phys.* **87**, pp. 1355–1360.
14. Teraoka, I., and Arnold, S. (2006). Enhancing the sensitivity of a whispering-gallery mode microsphere sensor by a high-refractive-index surface layer, *J. Opt. Soc. Am. B*, **23**, pp. 1434–1441.
15. Boriskina, S. V., and Reinhard, B. M. (2011). Adaptive on-chip control of nano-optical fields with optoplasmonic vortex nanogates, *Opt. Express*, **19**, pp. 22305–22315.
16. Smirnov, A. Y., Rashkeev, S. N., and Zagoskin, A. M. (2002). Polarization switching in optical microsphere resonator, *Appl. Phys. Lett.*, **80**, pp. 3503–3505.
17. Barber, P. W., and Chang, R. K. (1998). *Optical Effects Associated with Small Particles*. Singapore, World Scientific.
18. Teraoka, I., and Arnold, S. (2006). Theory of resonance shifts in TE and TM whispering gallery modes by nonradial perturbations for sensing applications, *J. Opt. Soc. Am. B*, **23**, pp. 1381–1389.
19. Arnold, S., Khoshsima, M., Teraoka, I., Holler, S., and Vollmer, F. (2003). Shift of whispering-gallery modes in microspheres by protein adsorption, *Opt. Lett.*, **28**, pp. 272–274.
20. Teraoka, I., Arnold, S., and Vollmer, F. (2003). Perturbation approach to resonance shifts of whispering-gallery modes in a dielectric microsphere as a probe of a surrounding medium, *J. Opt. Soc. Am. B*, **20**, pp. 1937–1946.
21. Ilchenko, V. S., Volikov, P. S., Velichansky, V. L., Treussart, F., Lefevre-Seguin, V., Raimond, J. M., and Haroche, S. (1998). Strain-tunable high-*Q* optical microsphere resonator, *Opt. Commun.*, **145**, pp. 86–90.
22. Tzeng, H.-M., Long, M. B., Chang, R. K., and Barber, P. W. (1985). Laser-induced shape distortions of flowing droplets deduced from morphology-dependent resonances in fluorescence spectra, *Opt. Lett.*, **10**, pp. 209–211.
23. Lai, H. M., Leung, P. T., Young, K., Barber, P. W., and Hill, S. C. (1990). Time-independent perturbation for leaking electromagnetic modes in open systems with application to resonances in microdroplets, *Phys. Rev. A*, **41**, pp. 5187–5198.
24. Sukhorukov, G. B., Fery, A., Brumen, M., and Möhwald, H. (2004). Physical chemistry of encapsulation and release, *Phys. Chem. Chem. Phys.*, **6**, pp. 4078–4089.

25. Lulevich, V. V., Andrienko, D., and Vinogradova, O. I. (2004). Elasticity of polyelectrolyte multilayer microcapsules, *J. Chem. Phys.*, **120**, pp. 3822–3826.

26. Ashkin, A. (2000). History of optical trapping and manipulation of small-neutral particle, atoms, and molecules, *IEEE J. Quantum Electron.*, **6**, pp. 841–856.

27. O´Dwyer, D. P., Phelan, C. F., Ballantine, K. E., Rakovich, Y. P., Lunney, J. G., and Donegan, J. F. (2010). Conical diffraction of linearly polarised light controls the angular position of a microscopic object, *Opt. Express*, **18**, pp. 27319–27326.

28. Molloy, J. E., and Padgett, M. J. (2002). Lights, action: optical tweezers, *Contempor. Phys.*, **43**, pp. 241–258.

29. Ren, K. F., Grehan, G., and Gouesbet, G. (1996). Prediction of reverse radiation pressure by generalized Lorentz–Mie theory, *Appl. Opt.*, **35**, pp. 2702–2710.

30. Mie, G. (1908). Beitrage zur Optik trüber Medien, speziell kolloidaler Metallösungen, *Ann. Phys.*, **25**, pp. 377–445.

31. Lebedeva, O. V., Kim, B.-S., Vasilev, K., and Vinogradova, O. I. (2005). Salt softening of polyelectrolyte multilayer microcapsules, *J. Colloid Interface Sci.*, **284**, pp. 455–462.

32. Kwok, A. S., and Chang, R. K. (1993). Stimulated resonance Raman scattering of Rhodamine 6G, *Opt. Lett.*, **18**, pp. 1703–1705.

33. Snow, J. B., Qian, S.-X., and Chang, R. K. (1985). Stimulated Raman scattering from individual water and ethanol droplets at morphology-dependent resonances, *Opt. Lett.*, **10**, pp. 37–39.

34. Symes, R., Sayer, R. M., and Reid, J. P. (2004). Cavity enhanced droplet spectroscopy: principles, perspectives and prospects, *Phys. Chem. Chem. Phys.*, **6**, pp. 474–487.

35. Rakovich, Y. P., Donegan, J. F., Gaponik, N., and Rogach, A. L. (2003). Raman scattering and anti-Stokes emission from a single spherical microcavity with a CdTe quantum dot monolayer, *Appl. Phys. Lett.*, **83**, pp. 2539–2541.

36. Gaponik, N., Rakovich, Y. P., Gerlach, M., Donegan, J. F., Savateeva, D., and Rogach, A. L. (2006). Whispering gallery modes in photoluminescence and Raman spectra of a spherical microcavity with CdTe quantum dots: anti-Stokes emission and interference effects, *Nanoscale Res. Lett.*, **1**, pp. 68–73.

37. Rakovich, Y. P., Gerlach, M., Donegan, J. F., Gaponik, N., and Rogach, A. L. (2005). Three-dimensional photon confinement in a spherical

microcavity with CdTe quantum dots: Raman spectroscopy, *Phys. E*, **26**, pp. 28–32.

38. Gorska, M., and Nazarewicz, W. (1974). Application of the random-element isodisplacement model to long-wavelength optical phonons in $CdSe_xTe_{1-x}$ mixed crystals, *Phys. Status Solidi B*, **65**, pp. 193–202.

39. Byrne, S. J., Corr, S. A., Rakovich, T. Y., Gun'ko, Y. K., Rakovich, Y. P., Donegan, J. F., Mitchell, S., and Volkov, Y. (2006). Optimisation of the synthesis and modification of CdTe quantum dots for enhanced live cell imaging, *J. Mater. Chem.*, **16**, pp. 2896–2902.

40. Baranov, A. V., Rakovich, Y. P., Donegan, J. F., Perova, T. S., Moore, R. A., Talapin, D. V., Rogach, A. L., Masumoto, Y., and Nabiev, I. (2003). Effect of ZnS shell thickness on the phonon spectra in CdSe quantum dots, *Phys. Rev. B*, **68**, pp. 165306–165301/165307.

41. Zhang, J.-Y., Wang, X.-Y., Xiao, M., Qu, L., and Peng, X. (2002). Lattice contraction in free-standing CdSe nanocrystals, *Appl. Phys. Lett.*, **81**, pp. 2076–2078.

42. Serpenguzel, A., Arnold, S., Griffel, G., and Lock, J. A. (1997). Enhanced coupling to microsphere resonances with optical fibers, *J. Opt. Soc. Am. B*, **14**, pp. 790–795.

43. Rakovich, Y. P., Yang, L., McCabe, E. M., Donegan, J. F., Perova, T., Moore, A., Gaponik, N., and Rogach, A. (2003). Whispering gallery mode emission from a composite system of CdTe nanocrystals and a spherical microcavity, *Semiconductor Sci. Technol.*, **18**, pp. 914–918.

44. Filonovich, S. A., Rakovich, Y. P., Vasilevskiy, M. I., Artemyev, M. V., Talapin, D. V., Rogach, A. L., Rolo, A. G., and Gomes, M. J. M. (2000). Probing the exciton density of states in semiconductor nanocrystals using integrated photoluminescence spectroscopy, *Chem. Monthly*, **133**, pp. 909–918.

45. Ausman, L. K., and Schatz, G. C. (2008). Whispering-gallery mode resonators: surface enhanced Raman scattering without plasmons, *J. Chem. Phys.* **129**, pp. 054704–054710.

46. Anderson, M. S. (2010). Nonplasmonic surface enhanced Raman spectroscopy using silica microspheres, *Appl. Phys. Lett.*, **97**, pp. 131116–131113.

47. Hoheisel, W., Colvin, V. L., Johnson, C. S., and Alivisatos, A. P. (1994). Thereshold for quasicontinuum absorption and reduced luminescence efficiency in CdSe nanocrystals., *J. Chem. Phys.*, **101**, pp. 8455–8460.

48. Rodrigues, P. A. M., Tamulaitis, G., Yu, P. Y., and Risbud, S. H. (1995). Size selective photoluminescence excitation spectroscopy in CdSe nanocrystals, *Solid State Commun.*, **94**, pp. 583–587.

49. Micic, O. I., Cheong, H. M., Fu, H., Zunger, A., Sprague, J. R., Mascarenhas, A., and Nozik, J. (1997). Size-dependent spectroscopy of InP quantum dots, *J. Phys. Chem. A.*, **101**, pp. 4904–4912.

50. Rakovich, Y. P., Walsh, L., Bradley, L., Donegan, J. F., Talapin, D., Rogach, A. L., and Eychmüller, A. (2002). Size selective photoluminescence excitation spectroscopy in CdTe quantum dots, *Proc. SPIE*, **4876**, pp. 432–437.

51. Joly, A. G., Chen, W., McCready, D. E., Malm, J.-O., and Bovin, J.-O. (2005). Upconversion luminescence of CdTe nanoparticles, *Phys. Rev. B*, **71**, pp. 165304–165301/165309.

52. Rakovich, Y. P., Filonovich, S. A., Gomes, M. J. M., Donegan, J. F., Talapin, D. V., Rogach, A. L., and Eychmüller, A. (2002). Anti-Stokes photoluminescence in II-VI colloidal nanocrystals, *Phys. Status Solidi B*, **229**, pp. 449–452.

53. Talapin, D. V., Haubold, S., Rogach, A. L., Kornowski, A., Haase, M., and Weller, H. (2001). A novel organometallic synthesis of highly luminescent CdTe nanocrystals, *J. Phys. Chem. A* **105**, pp. 2260–2263.

54. Downing, E., Hesselink, L., Ralston, J., and Macfarlane, R. (1996). A three-color, solid-state, three-dimensional display, *Science*, **273**, pp. 1185–1189.

55. Hehlen, M. P., Güdel, H. U., Shu, Q., and Rand, S. C. (1996). Cooperative optical bistability in the dimer system $Cs_3Y_2Br_9$:10% $Yb^{3+}$ *J. Chem. Phys.*, **104**, pp. 1232–1244.

56. Parthenopoulos, D. A., and Rentzepis, P. M. (1989). Three-dimensional optical storage memory, *Science*, **245**, pp. 843–845.

57. Said, A. A., Wamsley, C., Hagan, D. J., Van Stryland, E. W., Reinhardt, B. A., Roderer, P., and Dillard, A. G. (1994). Third- and fifth-order optical nonlinearities in organic materials, *Chem. Phys. Lett.*, **228**, pp. 646–650.

58. Cahalan, M. D., Parker, I., Wei, S. H., and Miller, M. J. (2002). Two-photon tissue imaging: seeing the immune system in a fresh light, *Nat. Rev. Immunol.*, **2**, pp. 872–880.

59. Pettit, D. L., Wang, S. S. H., Gee, K. R., and Augustine, G. J. (1997). Chemical two-photon uncaging: a novel approach to mapping glutamate receptors, *Neuron*, **19**, pp. 465–471.

60. Auzel, F. (2004). Upconversion and anti-Stokes processes with f and d ions in solids, *Chem. Rev.*, **104**, pp. 139–174.

61. Filonovich, S. A., Gomes, M. J. M., Rakovich, Y. P., Donegan, J. F., Talapin, D. V., Gaponik, N. P., Rogach, A. L., and Eychmüller, A. (2003). Up-

conversion luminescence in colloidal CdTe nanocrystals, *MRS Proc.*, **737**, pp. 157–162.

62. Rakovich, Y. P., Donegan, J. F., Filonovich, S. A., Gomes, M. J. M., Talapin, D. V., Rogach, A. L., and Eychmüller, A. A. (2003). Up-conversion luminescence *via* a below-gap state in CdSe/ZnS quantum dots, *Phys. E*, **17**, pp. 99–100.

63. Rakovich, Y. P., Gladyshchuk, A. A., Rusakov, K. I., Filonovich, S. A., Gomes, M. J. M., Talapin, D. V., Rogach, A. L., and Eychmüller, A. (2002). Anti-Stokes luminescence of cadmium telluride nanocrystals, *J. Appl. Spectroscop.*, **69**, pp. 444–449.

64. Rusakov, K. I., Gladyshchuk, A. A., Rakovich, Y. P., Donegan, J. F., Filonovich, S. A., Gomes, M. J. M., Talapin, D. V., Rogach, A. L., and Eychmüller, A. (2003). Control of efficiency of photon energy up conversion in CdSe/ZnS quantum dots, *Opt. Spectroscop.*, **94**, pp. 921–925.

65. Wang, X., Yu, W. W., Zhang, J., Aldana, J., Peng, X., and Xiao, M. (2003). Photoluminescence upconversion in colloidal CdTe quantum dots, *Phys. Rev. B*, **68**, pp. 125318–125311/125316.

66. Chylek, P., and Zhan, J. (1989). Interference structure of the Mie extinction cross section, *J. Opt. Soc. Am. A*, **6**, pp. 1846–1851.

67. Bayer, M., Gutbrod, T., Reithmaier, J. P., Forchel, A., Reinecke, T. L., Knipp, P. A., Dremin, A. A., and Kulakovskii, V. D. (1998). Optical modes in photonic molecules, *Phys. Rev. Lett.*, **81**, pp. 2582–2585.

68. Fuller, K. A. (1991). Optical resonances and two-sphere systems, *Appl. Opt.*, **30**, pp. 4716–4731.

69. Miyazaki, H., and Jimba, Y. (2000). *Ab initio* tight-binding description of morphology-dependent resonance in a bisphere, *Phys. Rev. B*, **62**, pp. 7976–7997.

70. Mukaiyama, T., Takeda, K., Miyazaki, H., Jimba, Y., and Kuwata-Gonokami, M. (1999). Tight-binding photonic molecule modes of resonant bispheres, *Phys. Rev. Lett.*, **82**, pp. 4623–4626.

71. Arnold, S., Ghaemi, A., Hendrie, P., and Fuller, K. A. (1994). Morphological resonances detected from a cluster of two microspheres, *Opt. Lett.*, **19**, pp. 156–158.

72. Hara, Y., Mukaiyama, T., Takeda, K., and Kuwata-Gonokami, M. (2003). Photonic molecule lasing, *Opt. Lett.*, **28**, pp. 2437–2439.

73. Möller, B. M., Woggon, U., Artemyev, M. V., and Wannemacher, R. (2004). Photonic molecules doped with semiconductor nanocrystals, *Phys. Rev. B*, **70**, pp. 115323–115321/115325.

74. Rakovich, Y. P., Donegan, J. F., Gerlach, M., Bradley, A. L., Connolly, T. M., Boland, J. J., Gaponik, N., and Rogach, A. (2004). Fine structure of coupled optical modes in photonic molecules, *Phys. Rev. A*, **70**, pp. 051801(R)–051801/051804.

75. Rakovich, Y. P., Gerlach, M., Bradley, A. L., Donegan, J. F., Connolly, T. M., Boland, J. J., Przyjalgowski, M. A., Ryder, A., Gaponik, N., and Rogach, A. L. (2004). Confined optical modes in small photonic molecules with semiconductor nanocrystals, *J. Appl. Phys.*, **96**, pp. 6761–6765.

76. Rusakov, K. I., Gladyshchuk, A. A., Rakovich, Y. P., Gerlach, M., Donegan, J. F., Connolly, T. M., Boland, J. J., Gaponik, N., and Rogach, A. L. (2005). Modification of photon states in photonic molecules with semiconductor nanocrystals, *Opt. Spectroscop.*, **99**, pp. 483–487.

77. Barnes, M. D., Mahurin, S. M., Mehta, A., Sumpter, B. G., and Noid, D. W. (2002). Three-dimensional photonic "molecules" from sequentially attached polymer-blend microparticles, *Phys. Rev. Lett.*, **88**, pp. 015508/015501–015504.

78. Srinivasarao, M., Collings, D., Philips, A., and Patel, S. (2001). Three-dimensionally ordered array of air bubbles in a polymer film, *Science*, **292**, pp. 79–83.

79. Rakovich, Y. P., Boland, J. J., and Donegan, J. F. (2005). Tunable photon lifetime in photonic molecules: a concept for delaying an optical signal, *Opt. Lett.*, **30**, pp. 2775–2777.

80. Harrison, M. T., Kershaw, S. V., Burt, M. G., Rogach, A. L., Kornowski, A., Eychmuller, A., and Weller, H. (2000). Colloidal nanocrystals for telecommunications. Complete coverage of the low-loss fiber windows by mercury telluride quantum dots, *Pure Appl. Chem.*, **72**, pp. 295–307.

81. Kershaw, S. V., Harrison, M., Rogach, A. L., and Kornowski, A. (2000). Development of IR-emitting colloidal II–VI quantum-dot materials, *IEEE J. Selected Top. Quantum Electron.*, **6**, pp. 534–543.

# Chapter 6

# Biological Applications of Cadmium Telluride Semiconductor Quantum Dots

**Anthony Le Cigne,[a] Alyona Sukhanova,[a,b] and Igor Nabiev[a,b]**

[a]*University of Reims Champagne-Ardenne, 51100 Reims, France*
[b]*Laboratory of Nano-Bioengineering, Moscow Engineering Physics Institute, 115409 Moscow, Russian Federation*

igor.nabiev@gmail.com

## 6.1 Introduction: Quantum Dots in Biology

Quantum dots (QDs) or nanocrystals are semiconductor crystals whose size in all three dimensions is smaller than the exciton Bohr radius of its constituent materials. These QDs are often composed of a periodic arrangement of atoms from groups II and VI (e.g. cadmium telluride, CdTe) or III and V of the periodic table. These size and composition parameters, resulting in nanometer-scale particles made up of 100–100,000 atoms, are the essential condition for the emergence of the unique properties of QDs compared with individual atoms or with their corresponding bulk material. Indeed, in such small materials, quantum mechanics describes the emergence of a phenomenon called the quantum confinement

*Cadmium Telluride Quantum Dots: Advances and Applications*
Edited by John F. Donegan and Yury P. Rakovich

ISBN 978-981-4316-05-7 (Hardcover), 978-981-4316-09-5 (eBook)
www.panstanford.com

effect, which occurs in the case of QDs in all three dimensions of space, in quantum wires in two spatial dimensions and in quantum wells in one dimension [1,2]. This phenomenon is the source of most of the outstanding optical properties of QDs.

The quantum confinement effect is at the origin of the size-tunable fluorescence emission wavelength of QDs, the property paving a way to numerous industrial and bio-optical applications [3,4]. The progress in chemical methods of synthesis has made possible the production of QDs of desired size, featuring a homogeneous size distribution and a quantum yield approaching 70% [5]. Hence, by selecting an adequate constituent semiconductor material and by tuning the synthesis process in order to reach a particular nanoparticle size, one can obtain a collection of efficiently emitting QDs having a fluorescence ranging from the near-ultraviolet (e.g. ZnSe QDs [6]) to the near-infrared (NIR) (e.g. CdTe QDs) or even into the mid-infrared region of the optical spectrum (Fig. 6.1).

Apart from these intrinsic and unique quantum confinement properties, many of the QDs photophysical properties are competitive or superior to the ones of organic dyes. First, QDs are characterized by very high molar extinction coefficients, often exceeding $10^6$ $M^{-1} \cdot cm^{-1}$ at the first excitonic absorption peak, and even much higher in the blue and UV regions of the optical spectrum [7]. Moreover, they absorb photons from a large range of wavelengths thanks to their wide and quasi-continuous UV-visible absorption spectra, allowing the excitation of different-sized QDs by a single light source [5], whereas organic dyes have a narrower absorption window, each of them needing to be excited with a specific wavelength. Second, QDs emit their fluorescence in the form of a narrow and symmetric emission peak, when emission peaks from organic dyes are usually broader and asymmetric, most of the time red-tailed (Fig. 6.1 shows Rhodamine 6G as an example of an organic dye). QDs are also brighter than organic dyes thanks to their efficient absorption, and an extreme photostability makes this brightness stable in time. QDs were found to be 100 times more resistant against photobleaching than Rhodamine 6G [8], and 380 and 3800 times more photoresistant than AlexaFluor and fluorescein isothiocyanate (FITC), respectively [9]. QDs fluorescence, being due to the decay of an exciton, also has a much longer lifetime compared with organic dyes, typically ranging from 5 ns to hundred of nanoseconds [10].

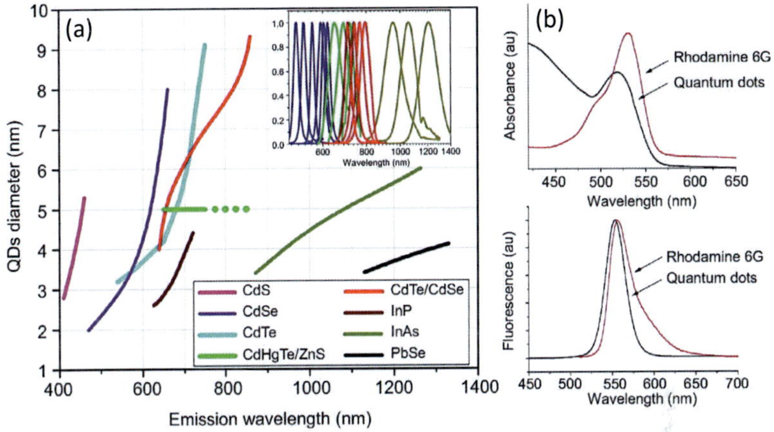


**Figure 6.1** (a) Emission maxima of QDs synthesized from various types of semiconductor materials (II–VI: CdS, CdSe, CdTe...; III–V: InP, InAs; IV–VI: PbSe) and varying in size. The curves represent experimental data from the literature and highlight the dependence between size, constituent material and wavelength of QDs fluorescence emission. Inset: Representative emission spectra of some of these QDs. Adapted from Ref. [75]. (b) Comparison between excitation (upper panel) and emission spectra (lower panel) of Rhodamine 6G (red) and CdSe QDs (black). The QDs excitation spectrum is wide and continuous, allowing their excitation at any wavelength shorter than ~530 nm. Rhodamine 6G has a much narrower absorption window. The QDs emission spectrum is narrow and symmetric while emission peak from Rhodamine 6G is broad and asymmetric. Adapted from Ref. [5].

These differences of the optical and physico-chemical properties between QDs and conventional organic fluorophores have proved to be of great importance in biotechnology [3,4]. The possibility to excite multicolor QDs with a single light source opens breakthrough perspectives in such a field as biomedical imaging as biomedical imaging, where signal multiplexing is of considerable importance, especially for applications to parallel and high-throughput analysis of multiple biological parameters or simultaneous detection of numerous biomarkers [11,12]. In addition, brightness and stability of QDs are of direct interest, since these properties are a necessary requirement for a fluorophore to be adapted to biological conditions, especially in single-molecule measurements [10,13,14], and in experiments of long-term tracking of different biological processes in living systems [15]. Finally, their long fluorescence lifetime allows for an enhanced contrast in time-gated biological imaging by reducing the contribution of the autofluorescence background commonly observed in biological conditions [16]. Owing to their size being similar to the one of biological macromolecules such as antibodies and nucleic acids, QDs' advantages as a biolabel have indeed found a use in several biotechnology fields, through an integration of nanotechnology and biology that constituted a major advance in the relatively new field of nanobiotechnology [3,4]. At the heart of this integration lies the ability to build a physical interface between biocompatible QDs and biological objects through bioconjugation [17].

QDs biocompatibility is achieved mostly through QDs' synthesis itself followed, when necessary, by modifications of their surface functionalities. Biocompatible QDs may be synthesized in water (e.g. CdTe QDs) or using a synthesis in organic solvents followed by different surface chemistry approaches to make them water soluble (e.g. CdSe QDs) (see Chapter 2). When made biocompatible, QDs exhibit multiple crucial properties, such as a long-term water solubility, presence of accessible functional groups in order to make bioconjugation possible, lack of perturbation of the conjugated biomolecule and finally biological innocuousness [18]. Among the various water solubilization techniques, ligand exchange, in which hydrophobic ligands on the QDs surface are replaced by bifunctional molecules like cysteine or cysteamine, with an end able to connect with QDs and the other one being both hydrophilic and reactive to biomolecules, is widely used. It is also possible to create a hydrophobic interaction between these hydrophobic ligands and the hydrophobic end of an amphiphilic polymer, the hydrophilic end being responsible for water solubility and providing functional groups for bioconjugation [5,8,18,19]. Silanization, a variant of ligand exchange in which QDs are coated with silica-based compounds in order to make them water soluble has also received attention, as did electrostatic attraction and nanobead-based conjugation [11,20]. Several techniques have been explored to link such biocompatible QDs with biomolecules. Among them can be cited passive adsorption, electrostatic interaction or standard covalent cross-linking approaches. The latter technique takes advantage of the hydrophilic surfactant shells reactive groups to realize the conjugation, through, for example, 1-ethyl-3-(3-dimethylaminopropyl)carbodiimide (EDC) coupling of carboxylic acid-functionalized QDs or bifunctional cross-linkers coupling with thiol-containing ligands such as 4-(*N*-maleimidomethyl)-cyclohexanecarboxylic acid *N*-hydroxysuccinimide ester (SMCC) for amine-functionalized QDs [17,21]. The avidin–biotin system has also been described in the context of QDs bioconjugation with biomolecules, even if covalent linkages are preferred when size is an important factor [22,23].

Hybrid materials made of biocompatible QDs and biomolecules result directly from these techniques and have been used in a very broad range of biological domains. Indeed, since the first use of QDs as biological labels by Bruchez *et al.* and Chan *et al.* [8,24], who managed to make CdSe QDs water soluble and to conjugate them with biomolecules, the number of biological studies involving QDs have increased dramatically [12,18,25,26]. Until now, the most prominent QDs for life science applications have been CdSe QDs surface-passivated with semiconductor material of wider bandgap, mostly ZnS, although CdTe, InP and InGaP QDs have also attracted some attention [27]. Other available cores such as ZnS, ZnSe, PbS, CdS and PbSe are not as common in biological applications [18]. Biological applications of QDs generally include sensing, imaging and the development of biomimetic systems. QDs have been used in the field of biosensing for immunoassays [28,29], nucleic acid detection [22], Förster resonance energy transfer (FRET)–based assays for the detection of various molecules such as proteases [30,31] and optical encoding for high-throughput screening [11]. QDs have also contributed to bioimaging through molecular and cellular labeling and cell tracking [15,32], single-molecule fluorescence methods [10,26] or immuno-histochemistry [33]. The involvement of QDs in biomimetic systems has also been part of QDs' role in nanobiotechnology [3,34,35].

In this chapter an overview of CdTe QDs biological applications is provided. After describing the CdTe QDs themselves with respect to life science applications, notably by comparing them with other biologically compatible QDs, various techniques of conjugation of these CdTe QDs with biomolecules will be presented. Applications of the resulting hybrid materials consisting of CdTe QDs and biomolecules will then be detailed for several domains of biology.

## 6.2 Cadmium Telluride Quantum Dots

### 6.2.1 Comparison with Other Biologically Compatible QDs

The factors that influence the compatibility of QDs with their use in life science, for example, water solubility, presence of accessible functional groups, lack of perturbation of the conjugated biomolecule and biological innocuousness [18], are mostly dependent on synthesis, surface passivation and surface chemistry of the QDs. To date, the majority of studies involving an application of QDs in life sciences were realized with core/shell CdSe/ZnS QDs and CdTe QDs [27,36]. CdSe/ZnS QDs are normally synthesized according to a method initially developed by Murray and coworkers which is based on the pyrolysis of an organometallic reagent at a relatively high-temperature, using surfactants such as tri-n-octylphosphine oxide (TOPO) or hexadecylamine (HDA) [37]. Although this procedure has been improved multiple times, for example, by the replacement of the precusor dimethyl cadmium by the more stable and less toxic CdO [38], it still inherently produces QDs that are hydrophobic and that have to be made water soluble to avoid their precipitation or aggregation in biological conditions. The subsequent water solubilization steps can sometimes be time consuming and difficult and can impair QDs' stability [39].

Other ways of efficiently producing QDs include direct aqueous synthesis, pioneered by Rajh *et al.* and Rogach *et al.* [40,41], then subsequently improved by numerous studies. Thiol-capped CdTe QDs, now commonly encountered, are among the nanoparticles produced using the latter method. Their synthesis in an aqueous medium, directly responsible for some interesting advantages over nanoparticles synthesized in organic solvents, has consequently been described as a possible useful alternative to organometallic synthesis, and they are indeed one of the main alternatives to CdSe/ZnS QDs for biological applications [36,42]. Nanoparticles synthesized that way are water soluble and can be coated with various thiol capping ligands, and consequently different functional groups. These thiols, which provide the stability in aqueous medium and functionality to the QDs, are normally thioglycolic acid (TGA) and 3-mercaptopropionic acid (MPA) but other thiol-containing compounds have been also used [36], such as L-cysteine [43] and *N*-acetylcysteine (NAC) [44]. These thiol-containing compounds are directly added to QDs during the synthesis process, and further steps of aqueous solubilization are optional. Depending on the need for a particular buffer or bioconjugation procedure, an adapted thiol-containing compound can be selected during the synthesis process. The resulting nanoparticles usually exhibit a good quantum yield and can be stable for several years [36]. It is worth mentioning that organometallically synthesized semiconductor nanoparticles, which have been commonly used to date, such as CdSe/ZnS, are inherently larger than aqueously synthesized CdTe QDs [27]. Indeed, apart from the presence of a shell made of semiconductor material of wider bandgap, which has been found to increase their quantum yield (QY) by reducing surface defects [45], supplementary shells have to be introduced on the surface of CdSe/ZnS QDs to make them water soluble. With aqueously synthesized CdTe QDs, this step is optional, and aqueously synthesized CdTe QDs are smaller than CdSe/ZnS even with a coating ligand such as TGA. This smaller size has been shown to be a direct factor of interest in biological applications. One can, for example, cite applications in which FRET from, or to QDs, is a crucial parameter. FRET, which is a very promising mean of detection for studying nanoscale interactions between biomolecules and a central parameter within nanophotonic devices, consists of the non-radiative transfer of energy between an energy donor and an acceptor when they are close to each other (a few nanometers) and when the fluorescence emission spectrum of the donor strongly overlaps the absorption spectrum of the acceptor [30,46,47] (see Chapter 4). FRET depends on $1/R^6$, where $R$ is the distance between the donor and the acceptor, and it has been found in some cases to be more efficient using small CdTe QDs than larger CdSe/ZnS QDs [3,35]. Some observations have also concluded that small CdTe QDs could access specific cellular compartments, such as the nucleus and nucleoli, while larger nanoparticles could not [48,49] (see Chapter 7).

These specific advantages of aqueously synthesized CdTe QDs are, nevertheless, counterbalanced by numerous flaws. Among them, the compatibility of CdTe QDs with common buffers has been described as depending on many unpredictable parameters [50]. Although the stability of CdTe QDs in some biological buffers can be reached under certain conditions, many authors have reported compatibility problems with biological buffers and media such as phosphate buffer saline buffer (PBS), Tris/Borate/EDTA buffer (TBE) [39] and human serum [3]. Mamedova *et al.*, in one of the first biological applications of CdTe QDs, reported that although it was possible to use as-synthesized CdTe QDs, a protective encapsulation such as silanization could be necessary to avoid the decomposition of the nanoparticles, their aggregation and nonspecific luminescence quenching in *in vitro* studies [43]. Silica shells have indeed been reported to enhance the stability of aqueously synthesized CdTe QDs in biological buffers such as PBS and TBE but provided significant decrease of nanoparticles fluorescence QY [39].

When *in vivo* applications are concerned, CdTe surface protection is necessary for two reasons: (1) to ensure QDs stability, since CdTe QDs have been found to be unstable when exposed to serum, cells or tissues [3], and (2) for preventing an exit of toxic ions from QDs to the media. ZnS shells, although they are responsible for an increase in size of CdSe/ZnS QDs, help to prevent the leakage of toxic $Cd^{2+}$ ions. Silica shells have also been reported as being a protective factor for cells submitted to Cd-containing nanoparticles. In a study carried out by Kirchner *et al.*, cytotoxicity of CdSe and CdSe/ZnS nanoparticles with different surfaces coatings such as MPA, silanization or polymer coating has been investigated [51]. Results showed that upon CdTe QDs' cellular uptake, acute cytotoxic effects due notably to the release of $Cd^{2+}$ ions were observed for nanoparticles with MPA and polymer coating, while silica-coated QDs showed no toxicity with 30-fold higher concentrations [51]. Further investigations of CdTe QDs toxicity have shown that their thinner ligand shell could also be a problem, since their degradation *in vivo* would lead to the exposition of biological structures to a bare semiconductor surface [52]. Intracellular concentrations of free $Cd^{2+}$ ions have also been found to be higher in cells exposed to CdTe QDs capped with various ligands than with CdSe/ZnS capped with cysteine [44], data that could also be linked to thinner coatings. Finally, the advantages of the small size of CdTe QDs and the simplicity of their aqueous synthesis are compromised by their lesser stability in biological media and a higher probability of toxic ions release in the environment. It is worth mentioning that a correct choice of nanoparticle should be always application driven, and smaller CdTe QDs present a serious alternative to bigger but more stable and better protected nanoparticles prepared according to organometallic routes [39].

As seen in Fig. 6.1, QDs made of CdTe allow for a fluorescence emission in the NIR region, which corresponds to the biological window of optical spectrum transmission by tissues and cells [36,53]. Indeed, living tissues exhibit a relatively low absorbance and scattering of light in the NIR, and *in vivo* fluorescence imaging requires deep photon penetration [54]. So-called type-II QDs, such as CdTe/CdSe core/shell QDs, which allow access to wavelengths that would not be available with a single (i.e. CdTe or CdSe) material are an interesting example of the use of CdTe within "composite" QDs, and is promising for applications in biomedical imaging in organisms, tissues and cells [53,55].

### 6.2.2 Methods of Quantum Dots Conjugation with Biomolecules

The first examples of QDs conjugation to biomolecules were demonstrated by Bruchez *et al.* and Chan *et al.* [8,24] for CdSe/ZnS QDs and further developed for QDs bioconjugates with proteins and peptides, oligonucleotides, polymers, carbohydrates, antibodies, and other biomolecules [3,18]. The explored methods of bioconjugation of CdSe/ZnS QDs include streptavidin–biotin affinity binding, amide bond formation, thioether bond formation or electrostatic interaction [56]. CdTe QDs have been covalently bioconjugated to proteins [43] and antibodies [29,57] and electrostatically linked to DNA *via* adaptors such as polymers [58,59] or directly electrostatically linked to proteins [35,60] and other molecules [61].

As previously stated, the surface chemistry of CdTe QDs can be tuned by replacing surface functionalities [36], what allows QDs to be obtained with variable surface charges and linkers on their surface. Some examples of covalent and non-covalent bioconjugation will be detailed below.

#### 6.2.2.1 Covalent bioconjugation

Covalent bioconjugation techniques notably include cross-linking of reactive groups present in the QDs organic coating, such as carboxylic acid and amino groups, with a relevant functionality in a biomolecule. Two methods have been mainly used for such covalent bioconjugation between CdTe QDs and biomolecules.

One of the first reports of a successful conjugation between CdTe QDs and biomolecules was described by Mamedova *et al.* [43], in a study dedicated to the preparation of albumin–(CdTe QDs) bioconjugates and concerned QDs with amino groups on their surface. L-cysteine-capped CdTe QDs were conjugated to bovine serum albumin (BSA) using the homobifunctional amine-reactive cross-linker glutaraldehyde (Fig. 6.2). In that reaction, one of the aldehyde groups of glutaraldehyde ($O{=}CH{-}(CH_2)_3{-}HC{=}O$) forms a Schiff base with the amino groups present at the surface of the QDs, *via* the L-cysteine. The remaining aldehyde group forms another Schiff base with BSA, *via* lysine amino acids.

CdTe QDs functionalized with carboxylic acid-containing molecules such as thioglycolic acid or 3-mercaptopropionic acid, which are the most often used ligands [36], can be conjugated to amino group-containing biomolecules using the *N*-hydroxysulfo-succinimide (sulfo-NHS) and/or 1-ethyl-3(3-dimethylaminopropyl) carbodiimide hydrochloride (EDC) reaction. These methods, which have been successfully used for the conjugation of CdTe QDs with proteins [62] and antibodies [29,57], consists in the action of EDC in the formation of a o-acylsourea reactive ester which will then react with sulfo-NHS to give a sulfo-NHS ester. The amino-containing biomolecules then reacts with the ester, providing the final amide derivative [21] (Fig. 6.2). It has been pointed out that this technique, which requires a low pH for the EDC/NHS coupling reaction, can decrease the QY of nanoparticles stabilized by thiols [57]. Better QY have been reported after the use of glutaraldehyde for conjugating L-cysteine-capped CdTe QDs [43].

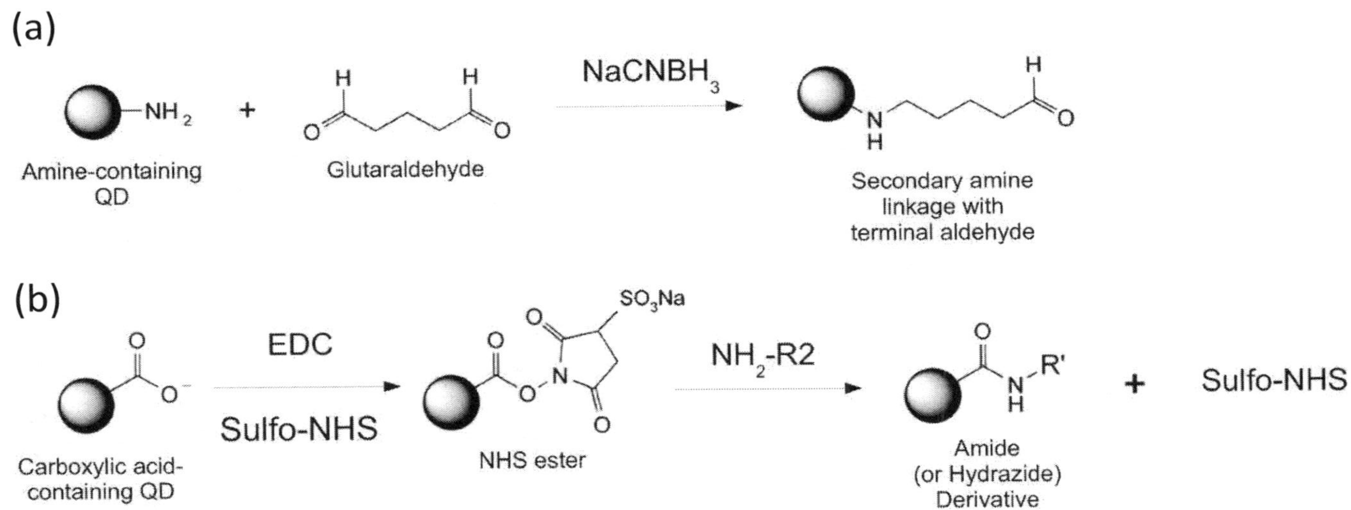


**Figure 6.2** (a) Bioconjugation using glutaraldehyde as a cross-linker. One of the aldehyde groups of glutaraldehyde forms a Schiff base with the surface amino groups, converted into stable secondary amine linkage by a reducing agent, while the other end is available for a reaction with a biomolecule. Adapted from Ref. [21]. (b) *N*-hydroxysulfo-succinimide (sulfo-NHS) and 1-ethyl-3(3-dimethylaminopropyl) carbodiimide hydrochloride (EDC) bioconjugation reaction. The surface reactive groups react with EDC and NHS to form a sulfo-NHS ester, which will react with an amino-containing biomolecule to form the final conjugate. Adapted from Ref. [21].

#### 6.2.2.2 Non-covalent coupling

Non-covalent coupling of QDs and biomolecules presumes the use of linkers between the nanoparticle and the biomolecule, in contrast to direct covalent conjugation by the mean of cross-linking agents. Direct electrostatic interaction can also be envisaged.

Applications of non-covalent coupling of biomolecules and CdTe QDs include notably DNA hybridization assays, in which a DNA probe has to be linked to QDs. Although covalent conjugation has been used in several studies for this purpose too, it can decrease the emission intensity of QDs and their stability [63,64], an obstacle that electrostatic coupling of QDs and DNA could overcome [58]. This technique has been used for CdSe/ZnS QDs [65] and CdTe QDs. CdTe QDs have been coupled with DNA through electrostatic interaction notably by the mean of a cationic polymer. Peng *et al.* [59] used the poly(diallyldimethylammonium chloride) (PDADMAC) cationic polymer to coat thiol-capped CdTe QDs in order to obtain positively charged QDs-polymers structures, able to interact electrostatically with a labeled DNA probe in a DNA hybridization assay based on FRET. In a similar experiment, Jiang *et al.* used another type of cationic polymer, poly[9,9-bis (3′-((*N*,*N*-dimethyl)-*N*-ethylammonium)propyl)-2,7-fluorene-alt-1,4-phenylene]dibromide (PDFD) to allow the dye-labeled DNA to interact with QDs [58]. These studies are detailed in the next section of this chapter (see Section 6.3.1.3).

## 6.3 Applications of Hybrid Materials Made from CdTe QDs and Biomolecules

To our knowledge, the first successful conjugation of CdTe QDs and a biomolecule was reported by Mamedova *et al.* [43], three years after the first bioconjugation of CdSe QDs realized by Bruchez *et al.* and Chan *et al.* [8,24]. In this proof-of-concept article, L-cysteine-capped CdTe QDs were synthesized in an aqueous solution and directly used for bioconjugation with BSA using the cross-linking agent glutaraldehyde, skipping the silica coating and/or ligand exchange steps. 1:1 BSA:CdTe conjugates were predominantly obtained, although 2:1 conjugates were also detected. Importantly, circular dichroism spectroscopy, the usual method for investigating proteins secondary structure, revealed that BSA was mostly intact when conjugated to QDs. Interestingly, a resonance energy transfer was also observed from BSA tryptophane moieties to QDs, thus acting as FRET acceptors in these conditions. The authors also pointed out the necessity of encapsulating the particles to protect them against decomposition and luminescence quenching. This study, by demonstrating the feasibility of CdTe QDs-biomolecules conjugation, paved the way for future design and application of such hybrid materials in various fields of biology.

This section is dedicated to the review of these applications in the fields of bioanalysis, in which QDs are used primarily as a mean of detection, and biomimetics, in which QDs are used as FRET donors.

## 6.3.1 Bioanalytical Tools

As previously stated, FRET is a powerful technique that allows for the detection of nanoscale interactions between optically coupled molecules [30]. QDs are well-fitted for applications in FRET-based assays since their emission wavelength is tunable and can then be adapted to the absorption spectrum of the acceptor. Importantly, FRET allows for the design of homogeneous assays, in which washing steps are not required anymore. FRET will consequently be largely involved in some of the applications presented further in the chapter.

### 6.3.1.1 Immunoassays

Antibodies are one of the affinity molecules which were linked to all types of QDs by various means such as cross-linking or by indirect streptavidin–biotin interaction of QDs coated with streptavidin and biotinylated antibodies [17]. Coating of CdTe QDs with antibodies was also carried out successfully, and one of the first proof-of-concept conjugation of CdTe QDs with antibodies is described below.

Wang *et al.* [57] used the sulfo-NHS/EDC method (see Section 6.2.2.1) to conjugate red-emitting and green-emitting thioglycolic acid-capped CdTe QDs to BSA and anti-BSA immunoglobulin (Ig) G, respectively. High-performance liquid chromatography (HPLC) analysis showed that no more than one nanoparticle were attached to either BSA, as previously reported by Mamedova *et al.* [43], or anti-BSA IgG. Protein secondary structure analysis using circular dichroism spectroscopy showed that BSA and anti-BSA IgG were not structurally affected by the conjugation to QDs. Binding affinity of anti-BSA IgG labeled with QDs ($QD_{green}$-IgG) was found to be 25–50% that of unlabeled IgG, whereas the affinity of a free polyclonal anti-BSA IgG for BSA labeled with QDs ($QD_{red}$-BSA) was not affected. On the formation of an immunocomplex between $QD_{green}$-IgG and $QD_{red}$-BSA, a FRET phenomenon occurred, which was expressed by a quenching of the green fluorescence of the IgG-conjugated QDs while increasing the red fluorescence of the BSA-conjugated QDs. This FRET effect was inhibited by addition of unlabeled BSA, since it replaced $QD_{red}$-BSA in the immunocomplex. Using this competitive FRET technique, the authors managed to detect BSA at a concentration as low as $10^{-8}$/M. It is worth mentioning that this technique does not need the binding and washing steps that are usually required by assays such as enzyme-linked immunosorbent assay (ELISA) and is therefore homogeneous.

#### 6.3.1.2 Applications of immunoassays employing QDs: detection of pathogens and chemicals

As demonstrated by Wang *et al.* [57], the bioconjugation of CdTe QDs with antibodies led to a functional detection device. Applications of such hybrid devices made of CdTe QDs and antibodies are described below.

Kuo *et al.* used the EDC/NHS bioconjugation method to link CdTe QDs to a polyclonal anti-*E. coli* antibody [29]. The results of cellular studies showed a colocalization between the red signal from ethidium bromide and the green signal from QDs-anti-*E. coli* antibodies hybrid probes. It was also shown that this colocalization could not be observed by using QDs alone as a probe or by using QDs-anti-*E. coli* probes with another bacteria, namely *Listeria monocytogenes*, confirming that the developed hybrid probe was specific in its detection of *E. coli* O157:H7 strain. These data thus confirmed previous results of Wang *et al.* [57], demonstrating that the covalent bioconjugation of an antibody to CdTe QDs can be realized without altering its recognition ability. It is worth mentioning that in this study, Kuo *et al.* used CdTe QDs coated with chemically reduced BSA (dBSA-CdTe QDs), as their previous findings suggested that such coating improved the chemical stability and QY of the QDs [66].

CdTe QDs have also been used by Chouhan *et al.* in designing a biosensor operating in a competitive binding assay for the pesticide methylparathion (MP) [62]. MPA-capped CdTe QDs were tagged with MP-BSA conjugate using the EDC/NHS technique. The bioconjugate was then used in a fluoroimmunoassay in which it competed with free MP for immobilized anti-MP IgY in an immuno-reactor column. The bioconjugate that was eluted out of the column was collected and its fluorescence measured. This fluorescence was found, as expected, to be directly proportional to the quantity of MP present in the sample, in a linear fashion for a range of concentrations between 1 and 0.1 $ng \cdot mL^{-1}$.

#### 6.3.1.3 Nucleic acids' detection

QDs have been previously suggested for use as fluorescent entities for DNA detection [67–69] and have been indeed used successfully for that purpose [70–72]. CdTe QDs have also received some attention in the context of DNA hybridization assays. In a study carried out by Peng *et al.*, a structure consisting of a thiol-capped CdTe QD rendered positively charged by wrapping with the cationic polymer PDADMAC (see Section 6.2.2.2) was electrostatically linked to a ssDNA probe labeled with Cyanine 3 (Cy3) (Fig. 6.3) [59]. Efficient FRET was obtained between the QD donor excited at 360 nm and Cy3 acceptor on the ssDNA probe, normally excited at 488 nm, through a partial spectral overlap of QDs fluorescence emission spectrum and Cy3 absorption spectrum. Upon addition of a complementary DNA strand in a concentration-dependent fashion, a decrease of Cy3 fluorescence was observed. The decrease in Cy3 fluorescence was confirmed to arise from a lower FRET efficiency from QDs to Cy3 when hybridization of complementary DNA strands occurred. However, upon addition of non-complementary DNA strands, a much slighter decrease was observed. This difference, which confirms the validity of the technique as a DNA hybridization assay, was explained by the authors as being due to a more rigid DNA duplex structure formed upon hybridization of a complementary DNA strand, which would increase the distance between the QD and Cy3, thus reducing FRET efficiency.

A possible contribution of an increase in repulsive forces between the QD and the duplex containing Cy3 was also considered.

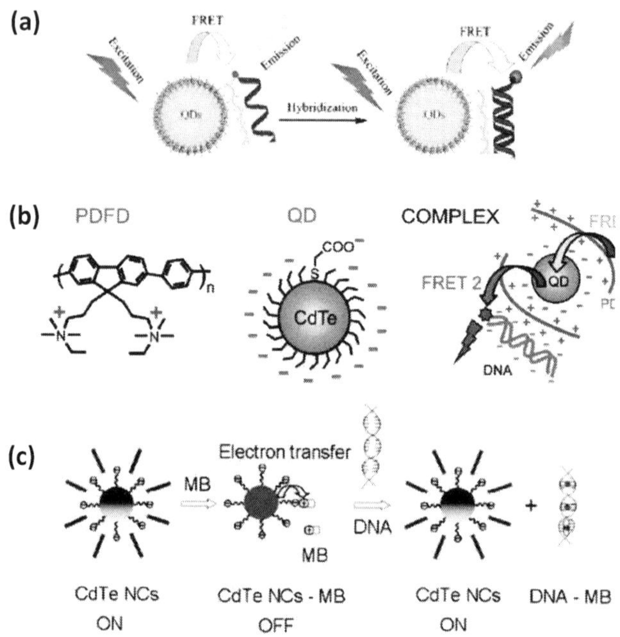


**Figure 6.3** Various applications of CdTe quantum dots (QDs) in nucleic acid detection. (a) Nanobiosensor built from an electrostatic interaction between a QD, a cationic polymer, and a Cy3-labeled DNA probe. Hybridization with complementary DNA increases the distance between QDs and Cy3, thus lowering fluorescence resonance energy transfer (FRET) from QDs to Cy3. A lesser emission intensity from Cy3 allows complementary DNA detection. Adapted from Ref. [59]. (b) Nanobiosensor built from an electrostatic interaction between a QD and a fluorescence-emitting cationic polymer. IRD700-labeled DNA probes in solution bind in a more effective way when in a DNA duplex form, in the presence of complementary DNA. FRET 1 improves the QDs fluorescence, while FRET 2 allows DNA duplex detection upon binding with the DNA duplex. Adapted from Ref. [58]. (c) Nanobiosensor built from direct electrostatic interaction between CdTe QDs and methylene blue (MB). Label-free DNA probes are in solution. QDs are quenched by electron transfer from QDs to MB. This quenching is ended upon formation of a DNA duplex, since MB is taken away from QDs by intercalating in this DNA duplex. Adapted from Ref. [61].

In another, but somewhat similar study, Jiang *et al.* designed a hybrid probe made of CdTe QDs and the cationic polymer PDFD (see Section 6.2.2.2) [58]. In this case, the cationic polymer rendered the QD positive, but also served as a light-harvesting antenna enhancing the QD luminescence through a FRET phenomenon (Fig. 6.3). In the solution, a dye-labeled ssDNA probe was present, but not interacting well with PDFD. However, upon addition of complementary DNA strands and formation of more negatively charged DNA duplexes, this dye-labeled duplex interacted with PDFD on the QDs surface. The dye label in the DNA duplex was found to be in close vicinity with the QDs and served as an acceptor in a detectable, quantitative FRET phenomenon thus allowing detection of the duplex formed by complementary DNA strands. Again, and as expected, non-complementary DNA did not cause a quenching of the QDs.

Both of the described approaches, although different, were based on the use of a cationic polymer for attraction. Direct electrostatic interaction approach has also found its place in nucleic acid detection. Indeed, a label-free DNA hybridization based on the direct electrostatic interaction between TGA-capped CdTe QDs and methylene blue (MB) has been developed by Shen *et al.* [61]. This assays used the fact that in QD–MB assemblies in solution, an electron transfer from CdTe QDs to MB occurs and causes a strong quenching of the QDs photoluminescence. MB has high affinity and binding constants to DNA duplexes, with which it interacts through intercalation between stacked DNA bases. In the developed DNA hybridization assay, unlabeled ssDNA was mixed with non-fluorescing QD–MB complexes. When complementary DNA strands were present in solution, DNA duplex formed and MB dissociated from the QD–MB complexes and intercalated within the dsDNA. Dissociation of QD–MB complexes produced strongly fluorescent individual QDs and this dissociation was used as indicative of the appearance of DNA duplexes. Addition of non-complementary DNA also caused a decrease of the quenching through, according to the authors, electrostatic interactions with MB, but to a much lower extent than with complementary DNA, thus validating the technique as a label-free DNA hybridization assay.

It is worth mentioning that one of the specific advantages of QDs is that each different DNA probe can be associated with QDs emitting a light of a precise wavelength, and which can be excited with a single light source. Consequently, in the case of a DNA hybridization assay using a specific DNA probe associated to a QD, a multiplexed, homogeneous assay can be carried out.

#### 6.3.1.4 Optical encoding

The first experiment involving QDs and optical encoding was carried out by Han *et al.* [11]. Multicolor hydrophobic ZnS-capped CdSe QDs were incorporated at controlled ratios into polymer microbeads and then further conjugated with DNA probes. In a DNA hybridization assay, the different types of DNA analytes could be detected and their associated code identified. "Bead-based" experiments with QDs are one of their most powerful applications, since their multiplexing capacities are in that case fully exploited. CdTe QDs have also been part of the development of optical encoding. Gaponik *et al.* have notably described a method of encapsulation of water-soluble CdTe QDs based on colloidal templating [73]. This technique consists in the construction of a double shell structure, a polyelectrolyte inner shell covered with an outer shell made of polyelectrolytes of opposed charge, around a melamine formaldehyde microsphere core. After dissolution of the core, the inner shell is decomposed, releasing the inner polymer inside the microcapsule. Microcapsules are then mixed with CdTe QDs, which penetrate in the microcapsule since it contains polyelectrolytes of opposed charge. The resulting fluorescent capsules provided a uniform and stable luminescence and were impermeable to any leakage of nanocrystals (Fig. 6.4). By incorporating QDs emitting at different wavelengths in the microcapsule, a reliable optical encoding has been obtained. Indeed, an almost linear relationship between red/green QDs concentration ratios and red/green fluorescence ratios was achieved with two-color tagged microspheres.

Using a layer-by-layer assembling strategy, Wang *et al.* [74] designed polystyrene beads coated with multiple layers of negatively charged thioglycolid acid-capped CdTe QD alternating with positively charged polyelectrolytes, and an outer layer of anti-IgG antibodies. After being characterized, these beads were used in an aggregation test, and fluorescence microscopy showed that upon a 5 min incubation with mouse IgG, the beads aggregated. When incubated with BSA, only individual beads were observable in solution. This indicates that the anti-IgG adsorbed on the beads are still biologically active and can recognize specific antigens, and that such beads could be used for targeting applications.

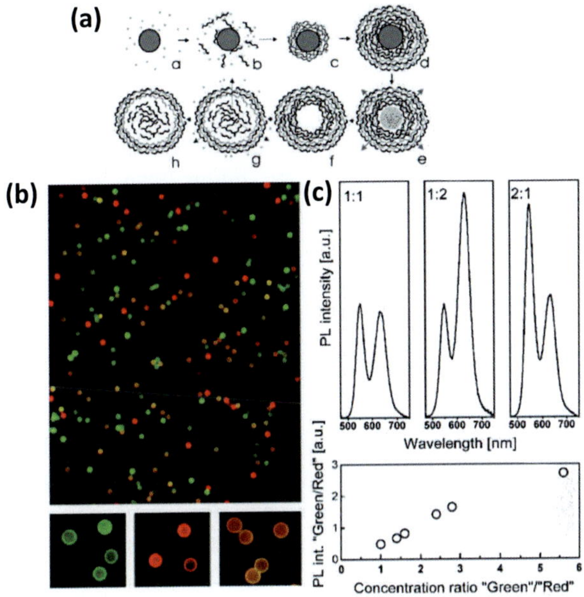


**Figure 6.4** Encapsulation of CdTe quantum dots (QDs) in microcapsules for multiplexed optical encoding. (a) Preparation of the polyelectrolyte-loaded microcapsules. A double shell structure is built around a core (a–d), then dissoluted (e). The inner shell is decomposed, releasing the polymer inside the microcapsule (f–h). CdTe QDs penetrate in the microcapsule now oppositely charged. (b) Fluorescence images of CdTe-loaded polymer microspheres. Upper panel shows a mixture consisting of single-color QDs and multicolor QDs. Lower panel shows three different types of loadings: green single-color QDs, red single-color QDs and a multicolor (five wavelengths). A true-color camera and a single light source were used. (c) Upper panel: fluorescence spectra of two-color tagged microspheres showing two separate peaks (548 and 633 nm) with controlled intensity ratios (1:1, 1:2, 2:1). Lower panel: relationship, in two-color microspheres, between the intensity of the green band (548 nm) and the relative concentration of the green-emitting CdTe QDs in the solution used for loading. The concentration of the red-emitting (633 nm) CdTe QDs was constant and their luminescence intensity was used for normalization. Adapted from Ref. [73].

#### 6.3.1.5 Bioimaging and biolabeling

QDs made from CdTe and particularly CdTe/CdSe core/shell type-II QDs are very well suited to bioimaging due to their ability to emit fluorescence in a "biological" window of the optical spectrum (see Section 6.2.1). Kim *et al.* used such type-II QDs, based on a CdTe core, to create an alternative to nuclear probes in sentinel lymph node (SLN) mapping [53]. Many cancers spread from their original site to regional lymph nodes, *via* the lymph channels. Removal of the regional lymph nodes can provide useful information about whether a chemotherapy is needed and in some case can be a direct factor of a cure. However, this operation is often the source of complications for the patient and is not a reliable clue of metastasis. SLN mapping is a technique that allows for the detection of the first lymph node that the cancer would encounter in case of metastasis. If no cancer cells are detected in the SLN, removal of regional lymph nodes is not necessary. SLN mapping is based on the injection of a radioactive substance near the original cancer, followed by the injection of a dye. The surgeon incises the skin where radioactivity is detected and then removes the SLN with the visual help of the dye. In the study by Kim *et al.*, NIR-fluorescent type II CdTe/CdSe (core/shell) QDs were used to realize the lymph node mapping (Fig. 6.5), providing an advantageous alternative

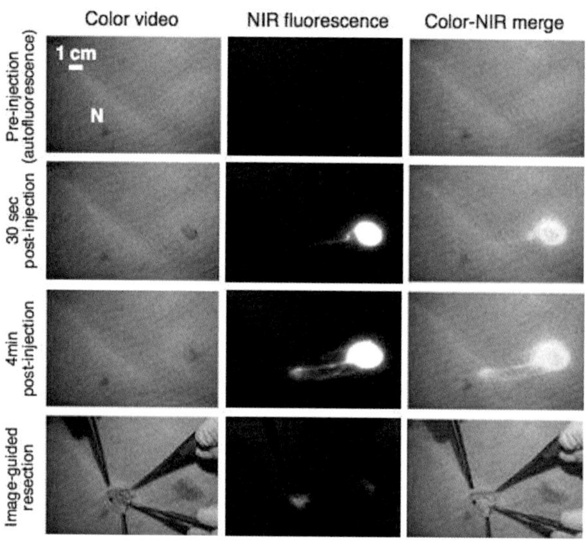


**Figure 6.5** Sentinel lymph node (SLN) mapping using type II CdTe/CdSe (core/shell) quantum dots (QDs). QDs are injected near the original cancer, and NIR fluorescence allows for the detection of the SLN in a straightforward way compared with previous techniques using classic dyes and radioactive tracers. Adapted from Ref. [53].

to nuclear isotopes. Indeed, first, it allowed the indentification of the lymph node and SLN optically and in real time, thus much more precisely than by using the radioactivity of nuclear probes. In addition, after resection, an optical feedback of the surgical site allows the surgeon for inspection and confirmation of the procedure completeness.

### 6.3.2 Biomimetic and Bioinspired Systems

Apart from their role as a fluorescent probe in bioanalytical tools, QDs have also been employed in biomimetic and bioinspired systems, and notably in semi-artificial bacterial photosynthetic systems. In nature, these structures can be found in various forms. The bacterium *Rhodobacter sphaeroides* uses protein-pigment antennas called light-harvesting antennas to absorb light and transfer excitons to a specialized structure known as reaction centers (RCs). This RC then dissociates excitons into their constituent carriers, which are then used in chemical transformations involved in the synthesis of highly energetic molecules that fuel the organism. Other bacterial organisms, such as *Halobacterium salinarum*, use a different photosynthetic system. In this case, the bacteria use a photochromic protein called bacteriorhodopsin (bR), which itself contains a retinal system able to absorb light. This process results in conformational changes in bR, which cause protons to be pumped through the so-called purple membrane (PM), creating a gradient that is then used by the ATPase to supply energy to cellular processes. Here, no light-harvesting antennas are used. RCs and retinals from bR could benefit from an association with QDs. Indeed, they are well suited to replace natural light-harvesting antennas and to complete the delicate retinal from bR for light absorption. Indeed, they are characterized by very high extinction coefficients, often exceeding extinction coefficients of natural light harvesting systems by orders of magnitude. Moreover, they absorb enormous quantities of solar energy in the UV-blue region, corresponding to high energy photons that photosensitive biological systems could not use without damage. Additionally, QDs are very resistant to photobleaching. Such semi-artificial photosynthetic devices have indeed been investigated.

Nabiev *et al.* [60] used CdTe QDs and CdSe/ZnS QDs as an additional light-harvesting antenna to improve the biological functions of RCs. These authors have electrostatically assembled

CdTe QDs with RCs purified from the light-harvesting complexes of *Rhodobacter sphaeroides* (Fig. 6.6), one of the most studied bacterial reaction centers. The results show that a FRET phenomenon takes place between QDs and RCs, and that this FRET phenomenon is more efficient with CdTe QDs than with CdSe QDs. This can be explained by the intrinsically smaller diameter of CdTe QDs, notably of their ligand shell. This FRET phenomenon was supported by various experiments. First, a quenching of

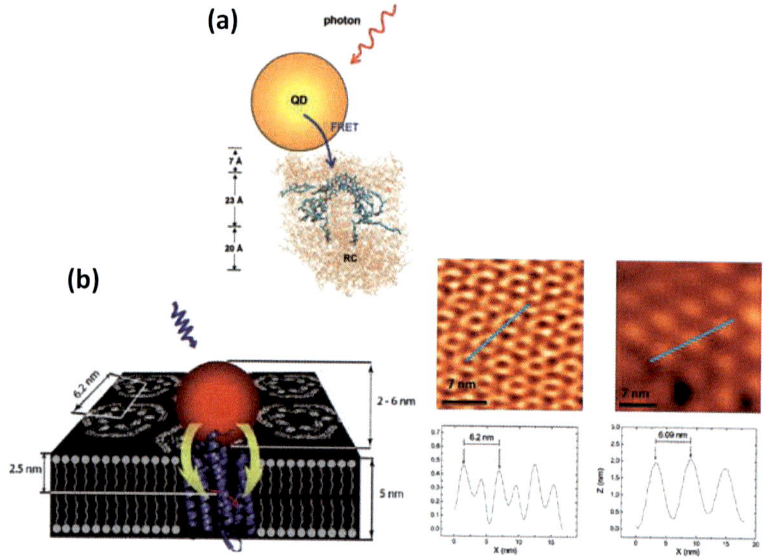


**Figure 6.6** Biomimetic systems based on CdTe QDs and biological molecules. (a) Biomimetic complex composed of a RC and a QD. FRET from QDs to RCs occurs, leading to an increase in the rate of generation of excitons in the RC thus to an improvement of the photosynthetic process. Diagram given to scale. Adapted from Ref. [60]. (b) Structural organization of PM-QD hybrid material. Left panel describes the FRET phenomenon taking place between a QD and the bR retinal (in purple). FRET from QD to retinal occurs, thus initiating photoisomerization, charge separation, bR photocycle, and proton pumping through the PM. This FRET results in a strong quenching of the PL of QD. Right panel: AFM images of bR within its natural PM (left) and assembled with QDs (right). AFM profiles scanned along the blue lines (bottom) indicate that the period of QDs organization upon assembling within the PM corresponds well to the period of bR trimer organization on the PM. Adapted from [35].

QDs fluorescence, directly dependent on the RCs:QDs ratio, was observed by the authors. In addition, the QDs fluorescence lifetime was shortened when assembled with RCs, in a correlated manner with quenching data. Fluorescence of the reduced RCs at 910 nm, due to the recombination of an exciton in the special pair upon reception of a photon, was found to be increased nearly three times for wavelengths corresponding to the solar spectrum when conjugated to QDs, a supplementary indication that QDs can indeed harvest light in a more efficient way than RCs alone.

In the same way as with RCs, bR has been investigated for its susceptibility to be functionally improved by nanotechnology and in particular by QDs. bR energy harvesting, which takes place without the help of any supplementary light-harvesting system in nature, has indeed been rendered more efficient by Rakovich *et al.*, in an experiment where CdTe QDs and CdSe/ZnS QDs and bR-containing PM were self-assembled (Fig. 6.6) [35]. Atomic force microscopy (AFM) images (Fig. 6.6) show QDs assembly with PM, following a pattern that corresponds to the natural organization of bR in the PM. From a functional point of view, as seen with RCs, an efficient quenching of QDs fluorescence and a shortening of their fluorescence lifetime occur when assembled with bR. Importantly, this quenching is not observed with white membrane (WM), a PM in which the bR retinal has been removed, strongly indicating a FRET phenomenon between QDs and bR. Again, size was an important factor, since quenching was slightly less efficient with CdTe QDs of larger diameter, and substantially reduced with larger CdSe/ZnS QDs. This FRET phenomenon was proved by the authors to cause an increase of bR efficiency in terms of proton pumping, their original function in nature. Proteoliposomes containing bR or QDs-bR hybrid complexes in their membrane, oriented in order to fill the vesicle interior with protons, were prepared. Measures of pH in the suspension of proteoliposomes showed an increase in pH upon illumination, corresponding to the pumping of protons from the solution to the proteoliposomes. The results showed a 25% increase of the bR photoresponse when it was embedded in QDs-bR hybrid complexes. Again, the size of QDs was found to be a crucial factor for the increase of bR functionality, since CdSe/ZnS QDs lead to a less important improvement of the photoresponse.

As demonstrated by these experiments, QDs, and especially CdTe QDs, are not only a promising tool for conceiving bioanalytical fluorescent probes, but can also be employed for light-harvesting and directional energy transfer, processing that are at the center of hybrid photovoltaics, artificial photosynthetic systems and optoelectronics [36].

## 6.4 Conclusion

The unique optical properties of QDs are of considerable interest for their integration in a wide range of nanodevices and are especially well suited to biological applications. The potential of QDs to complement or replace classic fluorescent dyes has attracted to them a lot of attention in the past decade, and advances in their synthesis, surface design and biocompatibilization have given birth to innovative nano-bio hybrid materials made of QDs and biomolecules. These advances have notably made possible the use of new types of QDs in this context, and CdTe QDs are one of the most successful examples of such alternatives to the more widely employed CdSe/ZnS QDs. Indeed, CdTe QDs have found a use in biosensing and bioimaging. They are in the center of QD-based biomimetic and bioinspired systems for light harvesting, and they also have proved their unique characteristics in terms of size and charge-selective cellular uptake that confirm their outstanding potential in bioimaging. Despite this success, which is a consequence of, among others, CdTe QDs easier synthesis, water solubility and reduced size, future works will have to investigate their specific flaws, such as their lack of stability in biological conditions, notably in common buffers and human serum, which is a direct obstacle to most of CdTe QDs *in vitro* and *in vivo* applications. Nevertheless, CdTe QDs hold great promise for different aspects of nanobiotechnological applications and will be very well represented in future applications of QDs in life sciences.

### Abbreviations

| | |
|---|---|
| AFM | atomic force microscopy |
| bR | bacteriorhodopsin |
| BSA | bovine serum albumin |

| | |
|---|---|
| Cy3 | cyanine 3 |
| DNA | deoxyribonucleic acid |
| dsDNA | double-stranded DNA |
| EDC | 1-Ethyl-3-(3-dimethylaminopropyl)carbodiimide |
| ELISA | enzyme-linked immunosorbent assay |
| FITC | fluorescein isothiocyanate |
| FRET | Förster resonance energy transfer |
| HDA | hexadecylamine |
| HPLC | high-performance liquid chromatography |
| Ig | immunoglobulin |
| MB | methylene blue |
| MP | methylparathion |
| MPA | 3-mercaptopropionic acid |
| NAC | *N*-Acetylcysteine |
| NIR | near-infrared |
| PBS | phosphate buffer saline |
| PDADMAC | poly(diallyldimethylammonium chloride) |
| PDFD | poly[9,9-bis(3′-((*N*,*N*-dimethyl)-*N*-ethylammonium) propyl)-2,7-fluorene-alt-1,4-phenylene]dibromide |
| PM | purple membrane |
| QD | quantum dot |
| QY | quantum yield |
| RC | reaction center |
| ROS | reactive oxygen species |
| SLN | sentinel lymph node |
| SMCC | 4-(*N*-maleimidomethyl)-cyclohexanecarboxylic acid *N*-hydroxysuccinimide ester |
| ssDNA | single-stranded DNA |
| sulfo-NHS | *N*-hydroxysulfo-succinimide |
| TBE | Tris/Borate/EDTA |
| TGA | thioglycolic acid |
| TOPO | tri-n-octylphosphine oxide |
| WM | white membrane |

## Acknowledgments

This study was partly supported by the European Commission through the FP7 cooperation program (grant number NMP-2009-4.0-3-246479 NAMDIATREAM) and the Ministry of Higher

Education and Science of the Russian Federation (grant number 11.G34.31.0050).

## References

1. Alivisatos, A. P. (1996). Semiconductor clusters, nanocrystals, and quantum dots, *Science*, **271**, pp. 933–937.
2. Nirmal, M., and Brus, L. (1999). Luminescence photophysics in semiconductor nanocrystals, *Acc. Chem. Res.*, **32**, pp. 407–414.
3. Nabiev, I., Sukhanova, A., Artemyev, M., and Oleinikov, V. (2008). *Colloidal Nanoparticles in Biotechnology*, ed. Elissari, A., Chapter 6 "Fluorescent Colloidal Particles as Detection Tools in Biotechnology Systems", (John Wiley, New York) pp. 133–168.
4. Sukhanova, A., and Nabiev, I. (2008). Fluorescent nanocrystal quantum dots as medical diagnostic tools, *Expert Opin. Med. Diagn.*, **2**, pp. 429–447.
5. Chan, W. C. W., Maxwell, D. J., Gao, X., Bailey, R. E., Han, M., and Nie, S. (2002). Luminescent quantum dots for multiplexed biological detection and imaging, *Curr. Opin. Biotechnol.*, **13**, pp. 40–46.
6. Hines, M. A., and Guyot-Sionnest, P. (1998). Bright UV-blue luminescent colloidal ZnSe nanocrystals, *J. Phys. Chem. B*, **102**, pp. 3655–3657.
7. Yu, W. W., Qu, L., Guo, W., and Peng, X. (2003). Experimental determination of the extinction coefficient of CdTe, CdSe, and CdS nanocrystals, *Chem. Mater.*, **15**, pp. 2854–2860.
8. Chan, W. C. W., and Nie, S. (1998). Quantum dot bioconjugates for ultrasensitive nonisotopic detection, *Science*, **281**, pp. 2016–2018.
9. Sukhanova, A., Devy, J., Venteo, L., Kaplan, H., Artemyev, M., Oleinikov, V., Klinov, D., Pluot, M., Cohen, J. H. M., and Nabiev, I. (2004). Biocompatible fluorescent nanocrystals for immunolabeling of membrane proteins and cells, *Anal. Biochem.*, **324**, pp. 60–67.
10. Dahan, M., Lévi, S., Luccardini, C., Rostaing, P., Riveau, B., and Triller, A. (2003). Diffusion dynamics of glycine receptors revealed by single-quantum dot tracking, *Science*, **302**, pp. 442–445.
11. Han, M., Gao, X., Su, J. Z., and Nie, S. (2001). Quantum-dot-tagged microbeads for multiplexed optical coding of biomolecules, *Nat. Biotech.*, **19**, pp. 631–635.
12. Medintz, I. L., Uyeda, H. T., Goldman, E. R., and Mattoussi, H. (2005). Quantum dot bioconjugates for imaging, labelling and sensing, *Nat. Mater.*, **4**, pp. 435–446.

13. Clarke, S., Pinaud, F., Beutel, O., You, C., Piehler, J., and Dahan, M. (2010). Covalent monofunctionalization of peptide-coated quantum dots for single-molecule assays, *Nano Lett.*, **10**, pp. 2147–2154.

14. Joo, C., Balci, H., Ishitsuka, Y., Buranachai, C., and Ha, T. (2008). Advances in single-molecule fluorescence methods for molecular biology, *Annu. Rev. Biochem.*, **77**, pp. 51–76.

15. Dubertret, B., Skourides, P., Norris, D. J., Noireaux, V., Brivanlou, A. H., and Libchaber, A. (2002). *In vivo* imaging of quantum dots encapsulated in phospholipid micelles, *Science*, **298**, pp. 1759–1762.

16. Dahan, M., Laurence, T., Pinaud, F., Chemla, D. S., Alivisatos, A. P., Sauer, M., and Weiss, S. (2001). Time-gated biological imaging by use of colloidal quantum dots, *Opt. Lett.*, **26**, pp. 825–827.

17. Mahmoud, W., Rousserie, G., Reveil, B., Tabary, T., Millot, J.-M., Artemyev, M., Oleinikov, V. A., Cohen, J. H. M., Nabiev, I., and Sukhanova, A. (2011). Advanced procedures for labeling of antibodies with quantum dots, *Anal. Biochem.*, **416**, pp. 180–185.

18. Mazumder, S., Dey, R., Mitra, M. K., Mukherjee, S., and Das, G. C. (2009). Review: biofunctionalized quantum dots in biology and medicine, *J. Nanomater.*, **2009**, pp. 1–17.

19. Yu, W. W., Chang, E., Drezek, R., and Colvin, V. L. (2006). Water-soluble quantum dots for biomedical applications, *Biochem. Biophys. Res. Commun.*, **348**, pp. 781–786.

20. Mattoussi, H., Mauro, J. M., Goldman, E. R., Anderson, G. P., Sundar, V. C., Mikulec, F. V., and Bawendi, M. G. (2000). Self-assembly of CdSe–ZnS quantum dot bioconjugates using an engineered recombinant protein, *J. Am. Chem. Soc.*, **122**, pp. 12142–12150.

21. Hermanson, G. T. (2008). *Bioconjugate Techniques*, 2nd Ed. (Elsevier Inc.).

22. Gerion, D., Chen, F., Kannan, B., Fu, A., Parak, W. J., Chen, D. J., Majumdar, A., and Alivisatos, A. P. (2003). Room-temperature single-nucleotide polymorphism and multiallele DNA detection using fluorescent nanocrystals and microarrays, *Anal. Chem.*, **75**, pp. 4766–4772.

23. Wu, X., Liu, H., Liu, J., Haley, K. N., Treadway, J. A., Larson, J. P., Ge, N., Peale, F., and Bruchez, M. P. (2003). Immunofluorescent labeling of cancer marker Her2 and other cellular targets with semiconductor quantum dots, *Nat. Biotech.*, **21**, pp. 41–46.

24. Bruchez, M., Moronne, M., Gin, P., Weiss, S., and Alivisatos, A. P. (1998). Semiconductor nanocrystals as fluorescent biological labels, *Science*, **281**, pp. 2013–2016.

25. Bailey, R. E., Smith, A. M., and Nie, S. (2004). Quantum dots in biology and medicine, *Physica E*, **25**, pp. 1–12.

26. Jamieson, T., Bakhshi, R., Petrova, D., Pocock, R., Imani, M., and Seifalian, A. M. (2007). Biological applications of quantum dots, *Biomaterials*, **28**, pp. 4717–4732.

27. Resch-Genger, U., Grabolle, M., Cavaliere-Jaricot, S., Nitschke, R., and Nann, T. (2008). Quantum dots versus organic dyes as fluorescent labels, *Nat. Methods*, **5**, pp. 763–775.

28. Goldman, E. R., Anderson, G. P., Tran, P. T., Mattoussi, H., Charles, P. T., and Mauro, J. M. (2002). Conjugation of luminescent quantum dots with antibodies using an engineered adaptor protein to provide new reagents for fluoroimmunoassays, *Anal. Chem.*, **74**, pp. 841–847.

29. Kuo, Y.-C., Wang, Q., Ruengruglikit, C., Yu, H., and Huang, Q. (2008). antibody-conjugated CdTe quantum dots for escherichia coli detection, *J. Phys. Chem. C*, **112**, pp. 4818–4824.

30. Medintz, I. L., Clapp, A. R., Brunel, F. M., Tiefenbrunn, T., Tetsuo Uyeda, H., Chang, E. L., Deschamps, J. R., Dawson, P. E., and Mattoussi, H. (2006). Proteolytic activity monitored by fluorescence resonance energy transfer through quantum-dot-peptide conjugates, *Nat. Mater.*, **5**, pp. 581–589.

31. Shi, L., De Paoli, V., Rosenzweig, N., and Rosenzweig, Z. (2006). Synthesis and application of quantum dots FRET-based protease sensors, *J. Am. Chem. Soc.*, **128**, pp. 10378–10379.

32. Smith, A. M., Ruan, G., Rhyner, M. N., and Nie, S. (2006). Engineering Luminescent quantum dots for *in vivo* molecular and cellular imaging, *Ann. Biomed. Eng.*, **34**, pp. 3–14.

33. Xing, Y., Chaudry, Q., Shen, C., Kong, K. Y., Zhau, H. E., Chung, L. W., Petros, J. A., O'Regan, R. M., Yezhelyev, M. V., Simons, J. W., Wang, M. D., and Nie, S. (2007). Bioconjugated quantum dots for multiplexed and quantitative immunohistochemistry, *Nat. Protoc.*, **2**, pp. 1152–1165.

34. Bouchonville, N., Molinari, M., Sukhanova, A., Artemyev, M., Oleinikov, V. A., Troyon, M., and Nabiev, I. (2011). Charge-controlled assembling of bacteriorhodopsin and semiconductor quantum dots for fluorescence resonance energy transfer-based nanophotonic applications, *Appl. Phys. Lett.*, **98**, pp. 013703.

35. Rakovich, A., Sukhanova, A., Bouchonville, N., Lukashev, E., Oleinikov, V., Artemyev, M., Lesnyak, V., Gaponik, N., Molinari, M., Troyon, M., Rakovich, Y. P., Donegan, J. F., and Nabiev, I. (2010). Resonance energy transfer improves the biological function of bacteriorhodopsin within

a hybrid material built from purple membranes and semiconductor quantum dots, *Nano Lett.*, **10**, pp. 2640–2648.

36. Gaponik, N., and Rogach, A. L. (2010). Thiol-capped CdTe nanocrystals: progress and perspectives of the related research fields, *Phys. Chem. Chem. Phys.*, **12**, pp. 8685–8693.

37. Murray, C. B., Norris, D. J., and Bawendi, M. G. (1993). Synthesis and characterization of nearly monodisperse CdE (E = sulfur, selenium, tellurium) semiconductor nanocrystallites, *J. Am. Chem. Soc.*, **115**, pp. 8706–8715.

38. Peng, Z. A., and Peng, X. (2001). Formation of high-quality CdTe, CdSe, and CdS nanocrystals using CdO as precursor, *J. Am. Chem. Soc.*, **123**, pp. 183–184.

39. Wolcott, A., Gerion, D., Visconte, M., Sun, J., Schwartzberg, A., Chen, S., and Zhang, J. Z. (2006). Silica-coated CdTe quantum dots functionalized with thiols for bioconjugation to IgG proteins, *J. Phys. Chem. B*, **110**, pp. 5779–5789.

40. Rajh, T., Micic, O. I., and Nozik, A. J. (1993). Synthesis and characterization of surface-modified colloidal cadmium telluride quantum dots, *J. Phys. Chem.*, **97**, pp. 11999–12003.

41. Rogach, A. L., Katsikas, L., Kornowski, A., Su, D., Eychmueller, A., and Weller, H. (1996). Synthesis and characterization of thiol-stabilized CdTe nanocrystals, *Ber. Bunsen-Ges.*, **100**, pp. 1772–1778.

42. Rogach, A. L., Franzl, T., Klar, T. A., Feldmann, J., Gaponik, N., Lesnyak, V., Shavel, A., Eychmüller, A., Rakovich, Y. P., and Donegan, J. F. (2007). Aqueous synthesis of thiol-capped CdTe nanocrystals: state-of-the-art, *J. Phys. Chem. C*, **111**, pp. 14628–14637.

43. Mamedova, N. N., Kotov, N. A., Rogach, A. L., and Studer, J. (2001). Albumin–CdTe nanoparticle bioconjugates: preparation, structure, and interunit energy transfer with antenna effect, *Nano Lett.*, **1**, pp. 281–286.

44. Cho, S. J., Maysinger, D., Jain, M., Röder, B., Hackbarth, S., and Winnik, F. M. (2007). Long-term exposure to CdTe quantum dots causes functional impairments in live cells, *Langmuir*, **23**, pp. 1974–1980.

45. Hines, M. A., and Guyot-Sionnest, P. (1996). Synthesis and characterization of strongly luminescing ZnS-capped CdSe nanocrystals, *J. Phys. Chem. B*, **100**, pp. 468–471.

46. Förster, T. (1948). Zwischenmolekulare Energiewanderung und Fluoreszenz, *Ann. Phys.*, **437**, pp. 55–75.

47. Lakowicz, J. R. (2006). *Principles of Fluorescence Spectroscopy*, 4$^{th}$ Ed. (Springer Verlag, New-York).

48. Nabiev, I., Mitchell, S., Davies, A., Williams, Y., Kelleher, D., Moore, R., Gun'ko, Y. K., Byrne, S., Rakovich, Y. P., Donegan, J. F., Sukhanova, A., Conroy, J., Cottell, D., Gaponik, N., Rogach, A., and Volkov, Y. (2007). Nonfunctionalized nanocrystals can exploit a cell's active transport machinery delivering them to specific nuclear and cytoplasmic compartments, *Nano Lett.*, **7**, pp. 3452–3461.

49. Williams, Y., Sukhanova, A., Nowostawska, M., Davies, A. M., Mitchell, S., Oleinikov, V., Gun'ko, Y., Nabiev, I., Kelleher, D., and Volkov, Y. (2009). Probing cell-type-specific intracellular nanoscale barriers Using size-tuned quantum dots, *Small*, **5**, pp. 2581–2588.

50. Boldt, K., Bruns, O. T., Gaponik, N., and Eychmüller, A. (2006). Comparative examination of the stability of semiconductor quantum dots in various biochemical buffers, *J. Phys. Chem. B*, **110**, pp. 1959–1963.

51. Kirchner, C., Liedl, T., Kudera, S., Pellegrino, T., Muñoz Javier, A., Gaub, H. E., Stölzle, S., Fertig, N., and Parak, W. J. (2005). Cytotoxicity of colloidal CdSe and CdSe/ZnS nanoparticles, *Nano Lett.*, **5**, pp. 331–338.

52. Lovrić, J., Cho, S. J., Winnik, F. M., and Maysinger, D. (2005). Unmodified cadmium telluride quantum dots induce reactive oxygen species formation leading to multiple organelle damage and cell death, *Chem. Biol.*, **12**, pp. 1227–1234.

53. Kim, S., Lim, Y. T., Soltesz, E. G., De Grand, A. M., Lee, J., Nakayama, A., Parker, J. A., Mihaljevic, T., Laurence, R. G., Dor, D. M., Cohn, L. H., Bawendi, M. G., and Frangioni, J. V. (2004). Near-infrared fluorescent type II quantum dots for sentinel lymph node mapping, *Nat. Biotech.*, **22**, pp. 93–97.

54. Lim, Y. T., Kim, S., Nakayama, A., Stott, N. E., Bawendi, M. G., and Frangioni, J. V. (2003). Selection of quantum dot wavelengths for biomedical assays and imaging, *Mol. Imaging*, **2**, pp. 50–64.

55. Kim, S., Fisher, B., Eisler, H.-J., and Bawendi, M. (2003). Type-II quantum dots: CdTe/CdSe(core/shell) and CdSe/ZnTe(core/shell) heterostructures, *J. Am. Chem. Soc.*, **125**, pp. 11466–11467.

56. Hild, W. A., Breunig, M., and Goepferich, A. (2008). Quantum dots—Nano-sized probes for the exploration of cellular and intracellular targeting, *Eur. J. Pharm. Biopharm.*, **68**, pp. 153–168.

57. Wang, S., Mamedova, N., Kotov, N. A., Chen, W., and Studer, J. (2002). Antigen/antibody immunocomplex from CdTe nanoparticle bioconjugates, *Nano Lett.*, **2**, pp. 817–822.

58. Jiang, G., Susha, A. S., Lutich, A. A., Stefani, F. D., Feldmann, J., and Rogach, A. L. (2009). Cascaded FRET in conjugated polymer/quantum dot/dye-labeled DNA complexes for DNA hybridization detection, *ACS Nano*, **3**, pp. 4127–4131.

59. Peng, H., Zhang, L., Kjällman, T. H. M., and Soeller, C. (2007). DNA hybridization detection with blue luminescent quantum dots and dye-labeled single-stranded DNA, *J. Am. Chem. Soc.*, **129**, pp. 3048–3049.

60. Nabiev, I., Rakovich, A., Sukhanova, A., Lukashev, E., Zagidullin, V., Pachenko, V., Rakovich, Y. P., Donegan, J. F., Rubin, A. B., and Govorov, A. O. (2010). Fluorescent quantum dots as artificial antennas for enhanced light harvesting and energy transfer to photosynthetic reaction centers, *Angew. Chem. Int. Ed.*, **49**, pp. 7217–7221.

61. Shen, J.-S., Yu, T., Xie, J.-W., and Jiang, Y.-B. (2009). Photoluminescence of CdTe nanocrystals modulated by methylene blue and DNA. A label-free luminescent signaling nanohybrid platform, *Phys. Chem. Chem. Phys.*, **11**, pp. 5062–5069.

62. Chouhan, R. S., Vinayaka, A. C., and Thakur, M. S. (2010). Thiol-stabilized luminescent CdTe quantum dot as biological fluorescent probe for sensitive detection of methyl parathion by a fluoroimmunochromatographic technique, *Anal. Bioanal. Chem.*, **397**, pp. 1467–1475.

63. Algar, W. R., and Krull, U. J. (2007). Towards multi-colour strategies for the detection of oligonucleotide hybridization using quantum dots as energy donors in fluorescence resonance energy transfer (FRET), *Anal. Chim. Acta*, **581**, pp. 193–201.

64. Zhou, D., Ying, L., Hong, X., Hall, E. A., Abell, C., and Klenerman, D. (2008). A compact functional quantum dot–DNA conjugate: preparation, hybridization, and specific label-free DNA detection, *Langmuir*, **24**, pp. 1659–1664.

65. Lee, J., Choi, Y., Kim, J., Park, E., and Song, R. (2009). Positively charged compact quantum dot–DNA complexes for detection of nucleic acids, *Chem. Phys. Chem.*, **10**, pp. 806–811.

66. Wang, Q., Kuo, Y., Wang, Y., Shin, G., Ruengruglikit, C., and Huang, Q. (2006). Luminescent properties of water-soluble denatured bovine serum albumin-coated CdTe quantum dots, *J. Phys. Chem. B*, **110**, pp. 16860–16866.

67. Algar, W. R., Massey, M., and Krull, U. J. (2009). The application of quantum dots, gold nanoparticles and molecular switches to optical nucleic-acid diagnostics, *Trends Anal. Chem.*, **28**, pp. 292–306.

68. Medintz, I. L., Clapp, A. R., Mattoussi, H., Goldman, E. R., Fisher, B., and Mauro, J. M. (2003). Self-assembled nanoscale biosensors based on quantum dot FRET donors, *Nat. Mater.*, **2**, pp. 630–638.

69. Rogach, A. L., Klar, T. A., Lupton, J. M., Meijerink, A., and Feldmann, J. (2009). Energy transfer with semiconductor nanocrystals, *J. Mater. Chem.*, **19**, pp. 1208–1221.

70. Mitchell, G. P., Mirkin, C. A., and Letsinger, R. L. (1999). Programmed assembly of DNA functionalized quantum dots, *J. Am. Chem. Soc.*, **121**, pp. 8122–8123.

71. Parak, W. J., Gerion, D., Zanchet, D., Woerz, A. S., Pellegrino, T., Micheel, C., Williams, S. C., Seitz, M., Bruehl, R. E., Bryant, Z., Bustamante, C., Bertozzi, C. R., and Alivisatos, A. P. (2002). Conjugation of DNA to silanized colloidal semiconductor nanocrystalline quantum dots, *Chem. Mater.*, **14**, pp. 2113–2119.

72. Pathak, S., Choi, S.-K., Arnheim, N., and Thompson, M. E. (2001). Hydroxylated quantum dots as luminescent probes for *in situ* hybridization, *J. Am. Chem. Soc.*, **123**, pp. 4103–4104.

73. Gaponik, N., Radtchenko, I. L., Sukhorukov, G. B., Weller, H., and Rogach, A. L. (2002). Toward encoding combinatorial libraries: charge-driven microencapsulation of semiconductor nanocrystals luminescing in the visible and near IR, *Adv. Mater.*, **14**, pp. 879–882.

74. Wang, D., Rogach, A. L., and Caruso, F. (2002). Semiconductor quantum dot-labeled microsphere bioconjugates prepared by stepwise self-assembly, *Nano Lett.*, **2**, pp. 857–861.

75. Michalet, X., Pinaud, F. F., Bentolila, L. A., Tsay, J. M., Doose, S., Li, J. J., Sundaresan, G., Wu, A. M., Gambhir, S. S., and Weiss, S. (2005). Quantum dots for live cells, *in vivo* imaging, and diagnostics, *Science*, **307**, pp. 538–544.

Chapter 7

# Biocompatibility and Cellular Localisation of CdTe Quantum Dots

**Yvonne Williams, Jennifer Conroy, and Yuri Volkov**

*NanoMedicine and Molecular Imaging Group,*
*School of Medicine and Institute of Molecular Medicine,*
*Trinity College Dublin, Ireland*

yvolkov@tcd.ie

*Space, the final frontier, to boldly go where no man has gone before....* This was the vision for the $20^{th}$ century, to expand our horizons beyond our planet and the known universe by exploring light years ($10^{15}$ m) into space. Yet all around us is a whole world waiting to be discovered at the nanoscale. Thus, the challenge of the $21^{st}$ century is to explore structures, systems and materials on the order of magnitude of $10^{-9}$ m as well as of $10^{9}$ m!

## 7.1 Introduction

The characteristics of nanoparticles, including their size on the same scale as many biomolecules, their large surface-to-volume ratio, their specific physical properties (strength, photoluminosity, magnetism) and their stability, render them as potential therapeutic

*Cadmium Telluride Quantum Dots: Advances and Applications*
Edited by John F. Donegan and Yury P. Rakovich

ISBN 978-981-4316-05-7 (Hardcover), 978-981-4316-09-5 (eBook)
www.panstanford.com

and diagnostic agents [1]. The ideal case scenario would be a nano-carrier that could be administered effectively, targeting a specific area, molecule or cell with minimal damage to surrounding tissue, and subsequently cleared, leaving no remains of residual material.

Initially, during the last decade, there has been significant optimism in using semiconductor nanocrystals (quantum dots) for biomedical applications. Thus, extensive research has been undertaken over recent years on the synthesis and characterisation of various types of quantum dots (QDs), particularly those with cadmium-containing cores [2–4]. While many studies favour QDs comprising a cadmium selinide core with a zinc sulphide shell (CdSe/ZnS) for use in the biological milieu, QDs with a core of cadmium telluride (CdTe) are advantageous as they can be synthesised directly in water as originally described by Weller's group [4], and therefore simpler, cheaper and safer to produce [3]. Aqueous synthesis also facilitates the production of much smaller QDs compared with their capped CdSe/ZnS counterparts (see Chapter 2).

A distinctive property of QDs is their ability to emit light of specific narrow wavelengths, depending on the size of the particle. The addition of a shell and/or hydrophilic surface coating enables the conjugation of various bioactive groups, including peptides, antibodies and drugs [5]. A wide range of biomolecules has been successfully conjugated to QDs with a view to their possible use as bioimaging agents [6,7]. However, it should be noted that the addition of such molecules might interact with the surface of QDs affecting their optical properties [8]. Capping agents, proteins and contaminants can also alter the biocompatibility of QDs [9].

Nevertheless, the properties of these "artificial atoms" make them very useful nanoprobes into the nanobarriers of the mammalian cell. By probing these barriers, novel methods of cellular mechanisms can be investigated in a manner as close as possible to natural processes, thereby ensuring more favourable outcomes. As with any exploration, the choice of tools is of paramount importance. Such studies require probes of similar size as the biomolecules being investigated, using instrumentation that is sensitive enough to resolve the finest detail at the subcellular level, yet powerful enough to gather sufficient data to yield robust and meaningful information.

Presently one of the most powerful tools available to cell and molecular biologists is fluorescent microscopy [10]. Unlike conventional microscopy, in which the area of interest might be masked among the surrounding tissue, object-specific fluorescent probes can be visualised standing out against a black background. Utilisation of fluorescent probes in tracking molecules within cells has made a hugely valuable contribution in cell biology studies. However, conventional fluorescent probes (e.g. DAPI, FITC, TRITC) are prone to bleaching, interfering with cell biology and having different absorption bands, thus making multiplexing difficult [11,12]. QDs, on the other hand, are proving to be very useful *in vitro* probes. QDs of different sizes are useful fluorescent tags as they can emit light with various wavelengths when excited by low energy light that can be absorbed harmlessly by living cells [1] and because of their narrow emission wavelengths are very much suitable for multiplexing assays [13]. Changing conditions during synthesis, such as temperature, duration and the addition of functional groups, can influence the resulting size and shape of QDs and thereby their emission wavelengths.

While examination of images using fluorescent microscopy by qualified researchers remains the gold standard of utilising fluorescent probes, it can be extremely time consuming and highly subjective, especially if a large number of multiplexed assays are required. Alongside the increased use of fluorescent probes, recent years have witnessed a corresponding development of high-throughput imaging and quantitative analysis using automated fluorescent microscopy methods [14]. High content analysis (HCA) refers to bioimaging assay technologies based on automated digital microscopy and image data mining. A further advantage is the presence of an environmentally controlled chamber in some models, thus facilitating live-cell imaging [15].

HCA is intrinsically multiparametric, as from the one assay, the number of cells and cellular organelles, shape, size and area of any labelled components, alongside with co-localisation of a number of differently labelled structures and their subcellular distribution, can all be objectively quantified. Moreover, multiple measurements can be made on individual cells or at a population level (Fig. 7.1).

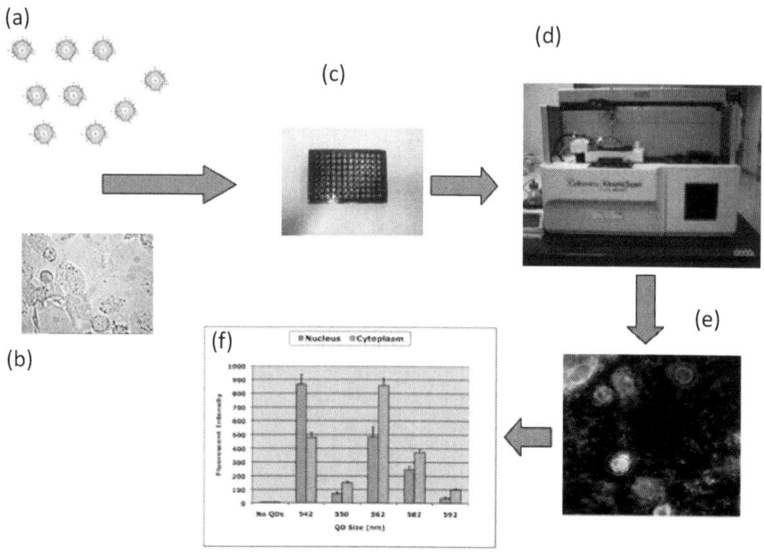


**Figure 7.1** Workflow demonstrating QD compartmentalisation using high content analysis. The compartmentalisation of size-tuned QDs (a) is examined in two distinctive cell lines, one epithelial, one macrophage (b), as these are representative of the most likely environmental exposure routes to nanoparticles in humans. The cells are seeded in 96 well microtitre plates (c) then fixed and permeabilised prior to the addition of the QDs, thus eliminating any cell membrane-associated effects due to active QD uptake mechanisms or to specificity of signalling routes in different cell types, but leaving intact the putative physical subcellular barriers. All assays are performed using a high content analysis (HCA) platform (d), resulting in high quality images to which user-defined algorithms could be applied (e) thereby getting robust data on large cell populations (f).

Investigations that can be performed using HCA include apoptosis, cytotoxicity, transcription factors, kinase activation, cell motility, cell shape changes, cell signalling events, intracellular trafficking, cell health measurement, cell viability and cell cycle analysis. It is possible to quantify these observations by refining algorithms during the image analysis process. Other technical advantages include the ability to use a wide variety of fluorescence indicators with excitation wavelengths from 350–700 nm. Because of auto-focusing and automatic exposure times, very rapid image acquisition can be achieved, thereby decreasing photo-damage and associated phototoxicity to the cells.

## 7.2 QDs as Fluorescent Probes into Cellular Barriers

As QDs are on a similar size scale as crucial biomolecules, including receptors, channels, ligands and nucleic acids, they can be adapted for various physiological effects [16]. Some of the many applications that have been reported include monitoring cellular receptor and intracellular trafficking, cellular movement, membrane dynamics, cellular development, evaluation of metastatic potential of malignant cells, multicolour cellular labelling for tracking, *in vivo* and tumour imaging and nanoparticle-mediated drug delivery [17]. However, there are a number of challenges which must be overcome as in any new applications for drug delivery, phototherapeutics, and *in vivo* or *in vitro* diagnostic imaging [18–20]. Methods must be devised to bypass the natural barriers that have evolved to prevent invasion by infectious agents. For drugs and phototherapeutics to be effective, they need to be able to target specific tissue, cellular and intracellular mechanisms. Any agent introduced into a living system should have minimal adverse effects to surrounding off-target cells prior, during, and post application, and eventually all traces of the active agent should be cleared from the system within a limited time frame.

Any agent, whether of natural or engineered origin, introduced into the living organism must first overcome the anatomical barrier consisting of the skin or mucous membranes, depending on whether the foreign agent is administered parentally, ingested or inhaled [21]. The primary barrier is a physical one consisting of epithelial cells, the function of which differs depending on the site to be protected. The upper layer of skin, comprising of squamous epithelium, is practically impermeable except by physical puncturing. In contrast, the columnar epithelial cells lining the mucosal layer of the respiratory and gastrointestinal tracts are designed to allow the regulated passage of gases, nutrients, and fluids between the exterior and the internal environment while remaining a protective layer for the underlying tissue and organs. Thus epithelial cells are important targets for drug or gene therapy due to their essential role in homeostasis [22]. There are specific receptors on the surface of these cells which can recognise certain molecules and carry them through the cell in a process known as transcytosis, alternatively some substances may pass between the cells [23].

The epithelial layer, however, is only an initial obstacle, followed by a phagocytic barrier comprising specialised cells for recognising and removing foreign matter into the reticuloendothelial system. These specialised cells, macrophages, are capable of internalising particles by a number of different endocytic mechanisms including phagocytosis. The internalised particle will be either processed or presented to the immune system *via* the lymph nodes, or alternatively, they end up in lysosomes designed to carry out the degradation and disposal of the particle *via* the excretory system [24].

There are a number of other functional barriers, including other divisions of the immune system, interlinked with the inflammatory response, a potent cocktail of cytokines, antibodies, and opsonin that can disable infectious agents, or speed up the process of their destruction [25]. Certain cells lining the mucosal surfaces have developed the ability to produce mucus that also prevents foreign agents from gaining access to the cell [26]. Similarly, molecular and ionic pumps can expel particles away from cells [27]. Furthermore, fluctuations in local physiological parameters such as temperature and pH can result in chemico-physical alterations to the particles, thus forming yet another barrier.

Nanoparticles larger than 6 nm will accumulate in the liver, lungs and the reticuloendothelial system (RES), while smaller ones are transported directly into the kidneys and bladder [28]. Other studies have demonstrated that nanoparticles with a hydrodynamic diameter (HD) of larger than 15 nm cannot be cleared by the kidneys, whereas anything under 5 nm can be cleared quite rapidly [29]. It also has been shown that nanoparticles of size 20–50 nm will travel straight to the lymph nodes, whereas larger particles (>500 nm) need to be taken up by dendritic cells and then brought to the lymph nodes for processing [30]. Any foreign particle has the potential to stimulate the immune response, either cell-mediated or antibody-mediated, depending on the particle size [31,32].

The cell itself is contained within a semi-permeable membrane, a phospholipid bilayer that has numerous endocytic mechanisms that allow the controlled passage of molecules entering and exiting the cell [33]. Furthermore, genetic information is tightly packaged into nuclei and mitochondria, organelles with a double membrane, each consisting of a hydrophobic phospholipid bilayer. Communication with the cytosol through these membranes is through highly complex pores that have specialised transport proteins controlling the movement of molecules into and out of the organelles [34].

The term endocytosis encompasses a number of different mechanisms that vary depending on the size, shape and charge of the endocytic cargo, the cell type, or on different transport conditions [33]. The simplest pathway is that of macropinocytosis, where the plasma membrane cups around very small molecules and absorbs them into the cytosol in a type of "drinking" process. Most other pathways mediate the transport of specific cargos through specific signals, on the cell surface resulting in the budding and pinching of the plasma membrane forming early endosomes. These can be divided into two main groups, clathrin dependent and clathrin independent, the latter including complex mechanisms some of which are not yet fully elucidated.

The mechanism of QD uptake by living cells is still open for discussion [35]; however, it is believed that non-receptor-mediated endocytosis is the most likely scenario [32]. It has been reported that the majority of functionalised QDs enter the cell *via* a clathrin-dependent pathway with a proportion entering *via* macropinocytosis [36]. It has also been shown that cells of macrophage lineage engulf nanoparticles within a few minutes of exposure indicating phagocytosis of aggregated particles [32], whereas epithelial cells are more likely to take the particles in by endocytosis in a size-dependent manner [37] (Fig. 7.2). Particles taken up by phagocytosis and contained in lysosomes can initiate size-dependent signalling that induces programmed cell death [38].

CdTe QDs are small enough to avoid the RES but can still be cleared from the bloodstream through the kidneys; hence, there is potential to develop QDs as *in vivo* probes. CdTe QDs of varying sizes have been employed as effective probes into the barriers

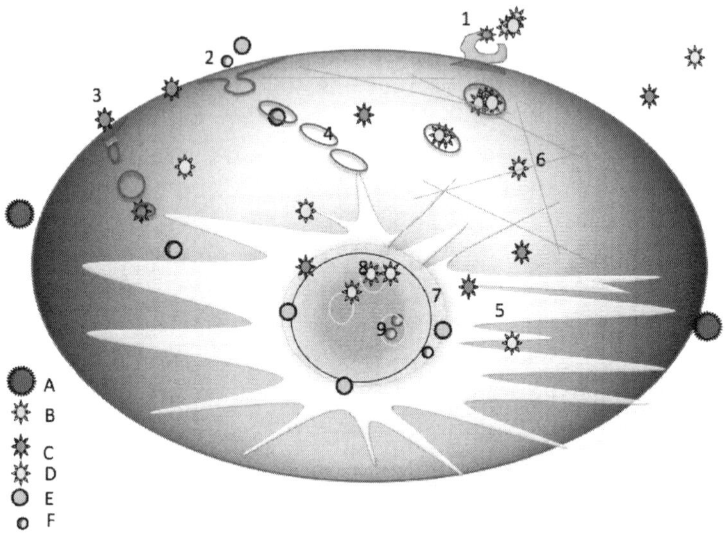


**Figure 7.2** QD entry and compartmentalisation in the cell. The typical eurkaryocytic cell consists of a hydrophobic plasma membrane enclosing the cytoplasm and nucleus, preventing entry to larger QDs (A). Endocytic mechanisms, the process of internalising macromolecules and particles through the plasma membrane, can be found in the majority of cell types. Phagocytes have receptors on their cell surface that recognise and attach to aggregated positively charged QDs (B), thus stimulating the formation of pseudopods (1). The QDs would then be processed into the lysosomes the acidic environment of which could possibly either alter the charge of the QDs or result in QD breakdown exposing the toxic core. Obviously this process could only occur when the QDs are actively taken up by living macrophages. The more usual pathway is that of macropinocytosis (2) where the plasma membrane cups around small negatively charged QDs (C) and absorbs them into the cytosol in a type of "drinking" process. Other pathways mediate the transport of QDs with biomolecules through specific receptors on the cell surface resulting in the budding and pinching of the plasma membrane (3) forming early endosomes (4). The cell is highly organised containing many functional sub compartments each limited by one or more membranes. The membranes are made mainly up of phospholipids, which form bilayers in aqueous solutions. The cytoplasm contains various organelles (Golgi apparatus, mitochondria, endoplasmic reticulum (ER), lysosomes), all of which are enclosed by hydrophobic membranes separating the various aqueous compartments. Depending on charge

and size QDs may localise in any of these compartments (5). The cytoskeleton microfilaments (actin and tubulin) play an important role not only in cellular structure, but also in the transport of QDs (6) and positioning of organelles within the cell. The nucleus is surrounded by a double membrane that forms the nuclear envelope (NE). The two layers fuse together forming the nuclear pore complex (NPC) that maintains an open communication between the cytosol and nucleus with mechanisms for signalling in both directions. Charge may be important in QD localisation with negatively charged particles more likely to accumulate in the perinuclear space (7). Supposedly there is free diffusion of molecules less than 9 nm, whereas active transport is required for molecules up to 35 nm. However, nanoparticles need to be less than 2.5 nm in size if entry into the nuclear space is to be observed (8, D). The exact cut-off depends on the lineage of the cell (i.e. macrophage or epithelial origin) (E). For nucleoli localisation the QD size is even smaller (9, F).

involved in cellular compartmentalisation. Cells of epithelial and phagocytic lineages grown in controlled laboratory conditions have been exposed to CdTe QDs and using the HCA technology mentioned above their size-dependent localisation within the cells have been tracked and quantified [39–41]. By partial permeabilisation of the cells with mild detergents prior to the addition of CdTe QDs, it is possible to avoid endocytic mechanisms and specific membrane signalling, demonstrating subcellular compartmentalisation of QDs determined by particle size, thereby extending previous observations on nucleo-cytoplasmic localisation of nanoparticles [40–42]. Furthermore, this compartmentalisation is specific to cell type with a wider size range of QDs observed in the cytoplasm and nucleus of macrophage cells than in epithelial cells. The rate the QDs were absorbed into the cells is also cell-type dependent.

The authors have demonstrated that CdTe QDs with a diameter of 2.1 nm ($\lambda_{em}$ 492 nm), and 3.1 nm ($\lambda_{em}$ 535.6 nm) can enter the nuclei and the smaller QDs locating to the nucleoli in both macrophage and epithelial cells. The larger 3.1 nm QDs are capable of entering the nucleoli of macrophages but not those of epithelial cells; 3.4 nm ($\lambda_{em}$ 580 nm) QDs are much slower to enter cells of both lineages and do not go beyond the cytoplasm. All the QDs investigated appear in epithelial cells more rapidly than in macrophages, with the cytoplasmic QDs disappearing out of the epithelial cells with time [41].

It would appear that the permeabilised cell membrane does not impose any significant barrier to the infiltration of nanoparticles into the cytoplasm, while the nuclear membrane in this case remains very size selective. The fluorescence from QDs displays a mesh-like pattern in the cytosol of macrophage cells indicating likely retention at the level of the endoplasmic reticulum (ER) [43]. As the ER is the site of protein assembling and packaging for secretion, it would be more extensive in macrophage cells than in epithelial cells, so it is not surprising that the QDs appeared as a fainter diffuse homogeneous pattern in the epithelial cells similar to that seen when inserting QDs by microinjection into the cytosol [44]. Accumulation of QDs at the perinuclear compartment within minutes has also been documented in this cell line by other authors [45,46]. This would indicate that there are components within the macrophage that bind strongly to the nanoparticles slowing down their entry into the nucleus. There is a suggestion that uptake of QDs may reach an equilibrium in epithelial cells [47] and that there may be a lower affinity for QDs within the epithelial cytoplasm in than observed in phagocytes.

For a particle to enter the nucleus, there may be a transient increase in nuclear pore size due to an accompanying signal-mediated transport [48]. Later studies have shown that passive diffusion of ions, metabolites and other small molecules can occur across the nucleocytoplasmic membrane reaching a Michaelis–Menton type equilibrium [47] while only molecules greater than 40 kiloDaltons (kD) require active transport mechanisms [49]. If QD are sufficiently small (i.e. ~2 nm or emitting in the green spectrum), they appear in the nucleus and eventually will locate at the nucleoli in all cell types examined. Also, the nuclear pore complexes are known to vary between different cell types [49], and this too may influence which QD enter the nuclei and which do not. Furthermore, there may be a correlation between nuclear pore size and nuclear envelope size with the larger pores occurring in smaller nuclear envelopes, thus implicating regulation of the flow of particles into and out of the nucleus [50].

### 7.2.1 Histone Targeting by CdTe Quantum Dots

Detailed investigation into TGA-capped CdTe 3 nm QDs confirmed that they have a distribution pattern within the cell similar to that displayed by histone proteins localising in the nucleus, and in particular the nucleoli. This affinity for the nucleoli is likely to be mediated by the interaction of the negatively charged QDs with the positively charged histones as previously described [40,51]. Using a combination of advanced technological approaches, including live cell confocal microscopy, fluorescent lifetime imaging (FLIM), spectroscopic methods, and zeta potential measurements, Conroy *et al.* demonstrated that unmodified CdTe preferentially bind to the positively charged core histone proteins. FLIM technology for the first time revealed that the localisation of these QDs to their ultimate nuclear and nucleolar destinations dramatically affected the QDs' photoluminescence lifetimes (PL), and thereby offers a sensitive readout for physical interactions between QDs and their intracellular macromolecular targets [51]. Another illustration of this variation in PL can be seen in Fig. 7.3, where epithelial cells have been exposed to CdSe QDs. The PL of the QDs is reduced in the nuclear compartment compared to the cytoplasmic regions due to their interaction with the intracellular macromolecules.

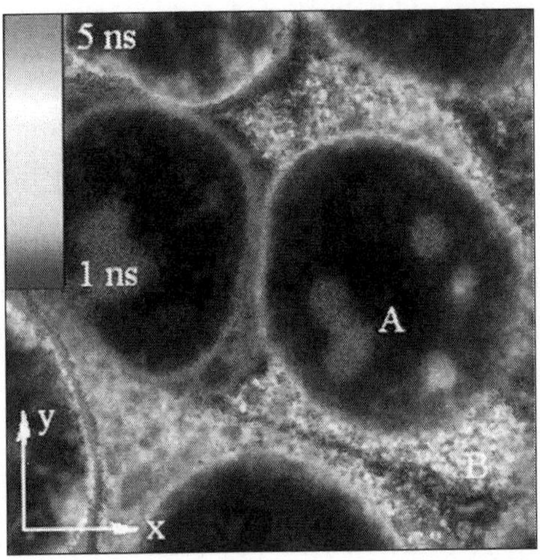


**Figure 7.3** Fluorescent lifetime imaging microscopy (FLIM). Fluorescent lifetime (FLIM) image of epithelial cells following exposure to CdSe QDs (A) nuclear region and (B) cytoplasmic region. A FLIM image is generated based on the photoluminescent lifetime of the fluorophore signal and not the intensity. Localisation of the QDs within the different cell compartments dramatically affects the QDs' photoluminescence lifetimes, which is due to the interaction of the QDs with the intercellular macromolecules.

The mode of QD uptake remains to be established (Fig. 7.2); however, it is certain that the transport of QDs within cells is energy dependent and not passive. Nabiev *et al.* showed that in order to reach the nucleus, CdTe QDs exploited the cells active microtubule transport route, whereas cells treated with anti-microtubule agent nocodazole lead to an accumulation of QDs in the cells cytoplasm rather than the nuclear region [40].

This affinity of CdTe QDs to the nuclei of cells is in many ways, however, a double-edged sword. Histones play a regulatory role in the normal cell cycle. This "tropism" of QDs to the histones during different stages of the cell cycle could in principle be exploited to determine the specific cell cycle stage of different cell types, at least in fixed cell samples. However, disruption of normal cell cycle by QDs in healthy cells could be catastrophic. Maysinger *et al.* showed that even low-level exposure to anionic CdTe QDs induced nuclear and mitochondrial disruption leading ultimately to genotoxic and epigenetic changes [5].

Alternatively, QDs can be used for direct manipulation at the level of subcellular components. Microinjection of QDs or mRNA-QD into the nuclei of Cos7 cells, described by Ishihama *et al.*, used oligo-QDs to track the movement of mRNA in the nucleus thus, allowing for single molecule tracking of the diffusion of mRNA in chromatin versus interchromatin regions of the nucleus [52].

Caution is paramount in the development and interpretation of cytotoxicity methods for assessing these nanomaterials, which are different from conventional drugs or chemicals used routinely in labs, and which most existing cytotoxicity assessment procedures and kits are developed for. It is therefore vital to use a broad range of techniques to measure the toxicity of QDs and other nanomaterials and realise that one single technique will not work for all.

## 7.3 CdTe QD Biocompatibility

Pharmaceutical companies are on a constant lookout for the latest "magic bullet" to cure infections, inflammatory diseases and, in particular, cancer [53]. Nanoparticles have been seen as possible candidates for this role since the start of the century. Furthermore, the European Technology Platform for Nanomedicine states that

the application of nanotechnology to medicine is the key for affordable efficient healthcare. Nanomedicine could result in the early diagnosis, rapid treatment, and effective follow-up care that would be available at a sustainable cost for all.
(ftp://ftp.cordis.europa.eu/pub/nanotechnology/docs/nanomedicine_bat_en.pdf)

Despite numerous assertions that *in vivo* experiments with QDs are well tolerated by the host [44,54,55], QD toxicity remains a serious concern. The main cause of QD cytotoxicity is believed to be the generation of reactive oxygen species caused by the release of $Cd^{2+}$ ions [9,56]. Oxygen radicals thus produced can damage cell organelles, rupture lysosomes, deform the nucleus and cause lipid peroxidation of the cell membrane resulting in apoptosis [5]. This inappropriate activation of the immune response can also result in further toxic effects [31]. While these same toxic effects could be harnessed to control pathological events, it should also be noted that the bioapplication of nanoparticles in general and QDs in particular raises the possibility of accidental release of these particles into the environment, where very little is known or understood about their potential hazards [57,58].

Unlike most pharmaceutical drugs, where knowledge of the chemical structure and interactive target sites is sufficient to investigate toxicity, the complex nature of QDs requires taking into account not only the chemical structure but also the physicochemical properties of the nanomaterial [16]. To ensure an adequate overview of toxicity, at least four parameters should be investigated, two qualitative and two quantitative [59]. On exposure to toxic compounds, cells can respond by variation in nuclear size and shape, increased cell permeability and increase in the number of lysosomes or change in mitochondrial potential. Decrease in cell numbers is also indicative of alteration in cell shape resulting in loss of adhesive properties and probable cell death. These biphasic responses can be seen in the same cell population as the cells in different parts of the cell cycle may manifest toxicity in a variety of ways [60]. Therefore it is essential that any multiparameter cytotoxicity assay is robust enough to show this data. HCA is particularly suitable for the investigation of cytotoxicity related to QD uptake [61].

Decreased toxic effects were seen with gelatine–QD nanocomposites exposed to live macrophages, suggesting that alteration to QD composition could eventually make them more biocompatible. CdTe QD–gelatine nanocomposites were prepared by the *in situ* synthesis of QDs in the presence of gelatine. The gelatine–QD composites readily pass through the cell membrane showing up in endosomes in the cytoplasm of macrophage cells. However, they are less likely to enter the nuclei. Compared with the original CdTe QDs, the gelatine-QD displayed much lower rates of toxicity, comparable to cells not exposed to QDs. The cadmium and TGA demonstrate the highest cell response rate in both of these tests and quickly induced toxic changes/cell death. These results show reduced toxicity for the CdTe QD–gelatine nanocomposites and also their potential as bioimaging agents [62] (Fig. 7.4).

Several investigations have shown that QD biocompatibility is determined not only by the cadmium-containing core [63] and surface functionality [64] but also by the nanoparticle size [42]. Coating the QDs with an epitaxial shell, such as ZnS, [63] or a protein [9] can increase biocompatibility and reduce leeching of toxic $Cd^{2+}$ ions from the metalloid core and also possibly reduce the role played by the capping groups through secondary interactions.

Thiol groups (sulphur-containing carboxyl molecules) are required for CdTe QDs as stabilisers to prevent breakdown and release of toxic Cd ions, enable water solubility and enhance the conjugation of biomolecules [3,39,65]. The addition of a thiol group can increase the diameter and diminish or enhance the quantum yield and allows for fine-tuning the surface charge. Any of these changes could have a significant effect on QD functionality and also on biocompatibility. A number of possible thiol stabilisers (thioglycolic acid (TGA), 3-mercaptopropionic acid (MPA), glutathione, L-cysteine, cysteamine) have been investigated for their chemical suitability and on how the uptake and localisation of CdTe QDs macrophages is affected.

TGA and MPA differ by one carbon atom and would be expected to have similar stabilising affects on CdTe QDs. Both TGA and MPA-QDs are taken up by macrophages in a similar manner and have similar sized dependent cellular localisation. The TGA-CdTe QDs are often the QD of choice as the technicalities of the synthesis is somewhat simpler than that of MPA QDs [39].

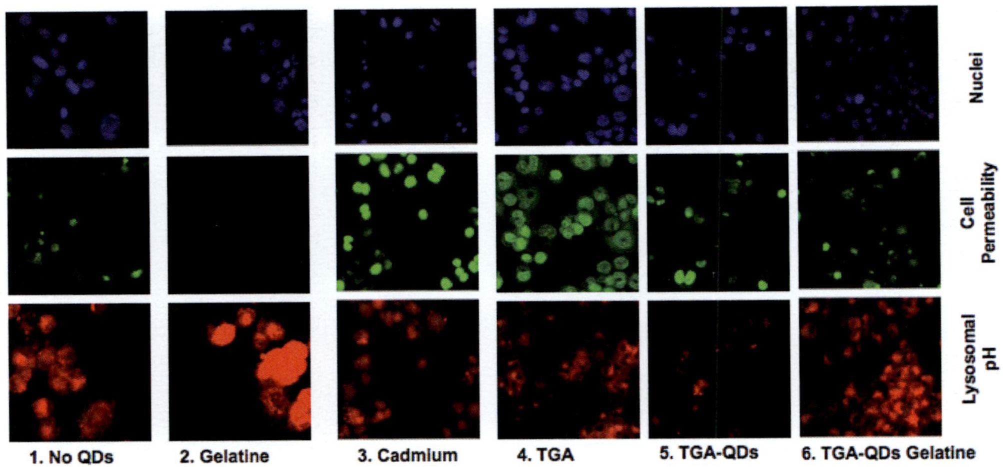


**Figure 7.4** Biocompatibility assay for surface modified CdTe QDs. Using three different dyes, many subtle changes in cell viability can be closely monitored. Increase in the sizes of the nucleus (stained with a blue dye) can indicate arrested cell cycle, while decreased size and increased fluorescent intensity is indicative of nuclear condensation and programmed cell death. The green dye for permeability is an inclusion dye and only enters cells with membrane damage. The red lysosomal dye is a weak base that locates within the acidic lysosomes, an increase in pH or a decrease in lysosomal mass due to toxicity results in a decrease in fluorescent intensity. Also the opposite can be true, i.e. an increase in lysosomal mass or decrease in pH is also an indication of toxicity. Healthy cells (1, 2) have round evenly stained nuclei, little or no permeability, and plenty of lysosomes. Components of quantum dots, cadmium (3) and TGA (4) have a more toxic affect on live cells that the QDs themselves (5), while the conjugation of gelatine, a protein polymer, to the QDs (6) decreases evidence of toxicity even further.

Glutathione, as a QD stabiliser, has a range of amino and carboxyl acid functionalities and a high quantum yield and therefore could be considered suitable for biocompatibility studies. Glutathione is found naturally in many organisms and coincidently has been found to detoxify cadmium in plant cells. Previous papers have shown successful internalisation of stable glutathione QDs [66] in methanol fixed cells, although there is no evidence of internalisation by living, probably due to size increase caused by aggregation. Glutathione-QDs are very unstable, lasting only a couple of weeks, before aggregating and falling out of solution.

Amino acids on the surface of QDs could act as natural ligands and enhance specificity for particular cells that would make amino acid QDs suitable candidates for drug delivery [67]. Cysteine is a naturally occurring amino acid, the only one containing a thiol group and thus is a natural source of sulphur for metabolism. It has an important biological role as an antioxidant and is a precursor of glutathione. L-cysteine-stabilised QDs accumulate at the cell surface of macrophages, regardless of the QD size [2]. It may be that the amino group binds to receptors on the cell surfaces. Cytotoxic analyses indicate that phagocytic cells exposed to cysteine-QDs show significant changes in nuclear size, cell permeability and lysosomal pH. It could be that the cysteine-stabilised QDs induce programmed cell death, a beneficial finding for the development of cytotoxic drugs.

Cysteamine is the simplest stable amino-thiol and is a degradation product of the amino acid cysteine. Also, it confers a positive charge onto the QDs resulting in them precipitating out of solution. Cytotoxicity studies reveal some cellular distress when the macrophage are exposed to cysteamine-QD or even cysteamine alone but change in cell number or nuclear size indicating mild cysteamine toxicity not necessarily related to QDs.

Other amino acids (AA) have also been investigated as suitable bio-conjugates on QDs. These include theonine, phenylalanine and tryptophan, all of which are essential amino acids and have been used extensively in the pharma industry. While phenylalanine and tryptophan are non-polar AA and contain an aromatic ring, theorine is polar and one of the few AA which contain an alcohol group. These were conjugated to both TGA-QDs and MPA-QDs. However, AA-MPA-QDs are very unstable and aggregate before cellular analysis could be performed. It has been suggested the presence of amino acids on QDs may cause increased aggregation and prevent dispersal of QDs in the natural environment [68]. Furthermore, a protein "corona" is believed to develop on the surface of nanoparticles when introduced into a biological milieu, enhancing uptake and decreasing toxicity [69].

Additionally, many of these findings have been replicated in epithelial cells lines, thus demonstrating consistency among different cell lineages, though as would be expected it takes longer for QDs to appear in epithelial cells and the range of different sizes is more limited (previously unpublished).

Although some thiol groups instigate QD aggregation, especially those that gave the nanoparticle a positive charge, others, particularly TGA and MPA enhance QD uptake by the cells and also improve photoluminescence. There is certainly evidence that there is some cytotoxicity, but nuclear changes normally associated with programmed cell death are rarely observed. While TGA and MPA prove to be the stabilisers of choice, the other thiols mentioned may still have a role in programmed cell death (cysteine) or in stabilising QDs utilised in post fixation staining (glutathione).

## 7.4 CdTe QDs as Possible Drug Delivery Agents

The ability to attach drugs to QDs could have beneficial implications, as the drug could be also tracked *in vivo* and therefore targeted. By combining QDs and amphipols (a nanomaterial that solubilises and delivers hydrophobic proteins into the cell membrane), siRNA has been delivered and released effectively into the cytosol while being imaged [70]. Using nanoparticles as a platform for drug delivery means that the solubility, biocompatibility and half-life depends on the engineered nanoparticle rather than the drug and therefore is possibly easier to manipulate [16]. QDs have been likened to micelles in their ability to carry drugs into cells, due to their ease of passing through the cell membrane [5]. However, surface–core modifications may be required. The addition of the Pep-1 peptide to the QD surface may act as a nuclear homing signal [71], while coating peptide-linked QDs with polyethylene glycol (PEG) slows down non-specific accumulation in the RES [54]. Gao *et al.* have shown that it is possible to target, image and possibly deliver drugs to cancer cells using QDs with a polymer that can be linked to tumour specific ligands [72].

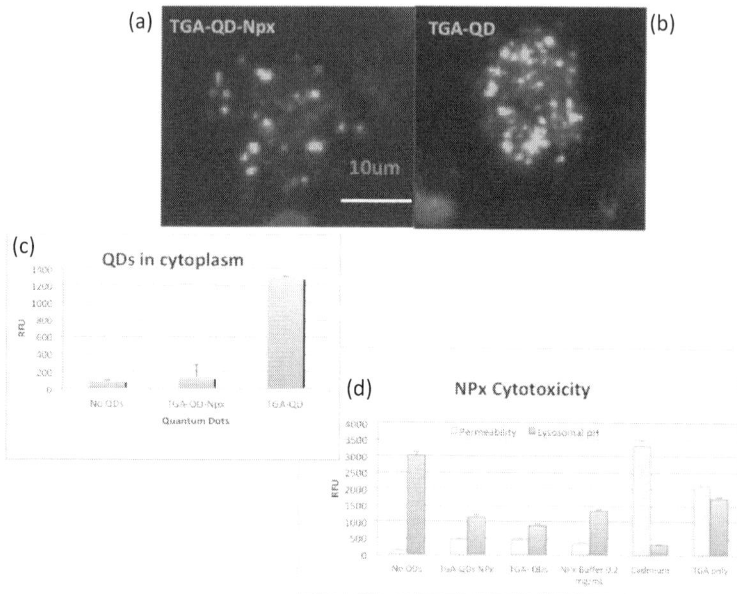


**Figure 7.5** Localisation of naproxen (NPx) conjugated QDs and non-conjugated QDs in the cytoplasm of macrophage cells. Activated macrophages are stimulated by liposaccharide (LPS) to ensure COX2 expression and then exposed to NPx CdTe nanocomposites (aA) and also to bare CdTe QDs (bB) for one hour in physiological conditions. While there appears to be a lower concentration of NPx Qds in the cells as indicated by intensity (c), there is no significant difference in localisation or toxicity (d) noted, thus demonstrating QDs as possible agents of drug delivery. Scale bar 10 microns.

Cyclooxygenase (COX2), an enzyme important in the production of inflammatory mediators, is found in activated macrophages. Naproxen, a non-steroidal anti-inflammatory drug, possesses well-defined anti-inflammatory properties as a COX2 inhibitor [73] and its fluorescent emission wavelength does not overlap with that of the CdTe QDs. Furthermore, the photophysical properties and biological behaviour of CdTe QDs ($\lambda_{em}$ 566 nm) conjugated to naproxen are not adversely affected. Conjugation can be confirmed by the final single spectrum of CdTe QDs. These nanocomposites have great potential as drug delivery and cellular imaging agents (Fig. 7.5). A similar approach could be used for development of other QD-drug conjugates, which would enable understanding and exploiting of the mechanistic pathways of drug delivery and help in diagnostic recognition and treatment of many diseases [74].

## 7.5 Summary

In this chapter, it has been demonstrated that CdTe QD can be very useful probes in the investigation of size-dependent functionalities of macrophages and epithelium layers, two very important targets and/or obstacles to drug delivery. Furthermore, as leaky blood vessels are a feature of malignant neoplasms, QD of various sizes can be used to target different tumours [75]. Moreover, in cancer treatment the pharmacology of QDs could be eventually harnessed and used in specific cell killing. Thus, CdTe QDs could be very powerful nanoprobes into the nanobarriers existing at the level of tissues, cells and subcellular structures.

The use of biocompatible polymers to reduce the toxicity of the semiconductor nanocrystals might be of great importance. Long-chain biopolymer molecules such as PEGs or gelatine can wrap the QDs, interacting with the capping groups and metal atoms on the surface. This could reduce the impact of the toxic CdTe-containing core. CdTe QDs are also suitable for chemical manipulation and can be capped with numerous thiol groups. They also are ideal for the addition of various biomolecules, which do not adversely affect the quantum yield or their localisation within the cells. While the issue of toxicity remains, coatings on QDs have been shown to help diminish the effects.

The addition of functional groups onto CdTe QDs can not only increase the range of biological applications but also alter their biocompatibility.

In conclusion, engineered nanoparticles are now being increasingly utilised in many biological applications, and in combination with high-powered microscopes embedded in automated platforms that can acquire and analyse large quantities of images, they provide new-generation tools for gaining invaluable information at the organ, tissue, cellular and subcellular level in an objective and accessible manner.

## References


1. Alivisatos AP, Gu W, Larabell C (2005). Quantum dots as cellular probes. *Annu Rev Biomed Eng*, **7**: 55–76.
2. Byrne S (2007). *Synthesis and Characterisation of CdTe QDs with Various Stabilisers*. PhD Thesis, School of Chemistry, Trinity College Dublin, Dublin.
3. Gaponik N, Talapin DV, Rogach AL, Hoppe K, Shevchenko EV, Kornowski A, Eychmuller A, Weller H (2002). Thiol-capping of CdTe nanocrystals: an alternative to organometallic synthetic routes. *J Phys Chem B*, **106**(29): 7177–7185.
4. Talapin DV, Haubold S, Rogach AL, Kornowski A, Haase M, Weller H (2001). A novel organometallic synthesis of highly luminescent CdTe nanocrystals. *J Phys Chem B*, **105**(12): 2260–2263.
5. Maysinger D, Lovric J, Eisenberg A, Savic R (2007). Fate of micelles and quantum dots in cells. *Eur J Pharm Biopharm*, **65**(3): 270–281.
6. Hoshino A, Manabe N, Fujioka K, Suzuki K, Yasuhara M, Yamamoto K (2007). Use of fluorescent quantum dot bioconjugates for cellular imaging of immune cells, cell organelle labeling, and nanomedicine: surface modification regulates biological function, including cytotoxicity. *J Artif Organs*, **10**(3): 149–157.
7. Smith AM, Duan H, Mohs AM, Nie S (2008). Bioconjugated quantum dots for *in vivo* molecular and cellular imaging. *Adv Drug Deliv Rev*, **60**(11): 1226–1240.
8. Zhang H, Zhou Z, Yang B, Gao M (2003) The influence of carboxyl groups on the photoluminescence of mercaptocarboxylic acid-stabilized CdTe nanoparticles. *J Phys Chem B*, **107**(1): 8–13.
9. Lovric J, Cho SJ, Winnik FM, Maysinger D (2005). Unmodified cadmium telluride quantum dots induce reactive oxygen species formation leading to multiple organelle damage and cell death. *Chem Biol*, **12**(11): 1227–1234.
10. Zorov DB, Kobrinsky E, Juhaszova M, Sollott SJ (2004). Examining intracellular organelle function using fluorescent probes: from animalcules to quantum dots. *Circ Res*, **95**(3): 239–252.
11. Medintz IL, Uyeda HT, Goldman ER, Mattoussi H (2005). Quantum dot bioconjugates for imaging, labelling and sensing. *Nat Mater*, **4**(6): 435–446.
12. Sukhanova A, Nabiev I (2008). Fluorescent nanocrystal quantum dots as medical diagnostic tools. *Expert Opin Med Diagn*, **2**(4): 429–447.

13. Sukhanova A, Nabiev I (2008). Fluorescent nanocrystal-encoded microbeads for multiplexed cancer imaging and diagnosis. *Crit Rev Oncol Hematol*, **68**(1): 39–59.

14. Pepperkok R, Ellenberg J (2006). High-throughput fluorescence microscopy for systems biology. *Nat Rev Mol Cell Biol*, **7**(9): 690–696.

15. Abraham VC, Taylor DL, Haskins JR (2004). High content screening applied to large-scale cell biology. *Trends Biotechnol*, **22**(1): 15–22.

16. McNeil SE (2009). Nanoparticle therapeutics: a personal perspective. *Wiley Interdiscip Rev Nanomed Nanobiotechnol*, **1**(3): 264–271.

17. Delehanty JB, Mattoussi H, Medintz IL (2009). Delivering quantum dots into cells: strategies, progress and remaining issues. *Anal Bioanal Chem*, **393**(4): 1091–1105.

18. Giepmans BN, Adams SR, Ellisman MH, Tsien RY (2006). The fluorescent toolbox for assessing protein location and function. *Science*, **312**(5771): 217–224.

19. Moghimi SM, Hunter AC, Murray JC (2001). Long-circulating and target-specific nanoparticles: theory to practice. *Pharmacol Rev*, **53**(2): 283–318.

20. Torchilin VP (2006). Recent approaches to intracellular delivery of drugs and DNA and organelle targeting. *Annu Rev Biomed Eng*, **8**: 343–375.

21. Stern ST, McNeil SE (2008). Nanotechnology safety concerns revisited. *Toxicol Sci*, **101**(1): 4–21.

22. Davda J, Labhasetwar V (2002). Characterization of nanoparticle uptake by endothelial cells. *Int J Pharm*, **233**(1–2): 51–59.

23. Stenbeck G, Horton MA (2004). Endocytic trafficking in actively resorbing osteoclasts. *J Cell Sci*, **117**(Pt 6): 827–836.

24. Desjardins M (2003). ER-mediated phagocytosis: a new membrane for new functions. *Nat Rev Immunol*, **3**(4): 280–291.

25. Han J, Ulevitch RJ (2005). Limiting inflammatory responses during activation of innate immunity. *Nat Immunol*, **6**(12): 1198–1205.

26. Behrens I, Pena AI, Alonso MJ, Kissel T (2002). Comparative uptake studies of bioadhesive and non-bioadhesive nanoparticles in human intestinal cell lines and rats: the effect of mucus on particle adsorption and transport. *Pharm Res*, **19**(8): 1185–1193.

27. Sanhai WR, Sakamoto JH, Canady R, Ferrari M (2008). Seven challenges for nanomedicine. *Nat Nano*, **3**(5): 242–244.

28. Ferrari M (2008). Nanogeometry: beyond drug delivery. *Nat Nano*, **3**(3): 131.

29. Soo Choi H, Liu W, Misra P, Tanaka E, Zimmer JP, Itty Ipe B, Bawendi MG, Frangioni JV (2007). Renal clearance of quantum dots. *Nat Biotechnol*, **25**(10): 1165–1170.

30. Manolova V, Flace A, Bauer M, Schwarz K, Saudan P, Bachmann MF (2008). Nanoparticles target distinct dendritic cell populations according to their size. *Eur J Immunol*, **38**(5): 1404–1413.

31. Dobrovolskaia MA, Germolec DR, Weaver JL (2009). Evaluation of nanoparticle immunotoxicity. *Nat Nano*, **4**(7): 411–414.

32. Xiang SD, Scholzen A, Minigo G, David C, Apostolopoulos V, Mottram PL, Plebanski M (2006). Pathogen recognition and development of particulate vaccines: does size matter? *Methods*, **40**(1): 1–9.

33. Mayor S, Pagano RE (2007) Pathways of clathrin-independent endocytosis. *Nat Rev Mol Cell Biol*, **8**(8): 603–612.

34. Mazzanti M, Bustamante JO, Oberleithner H (2001). Electrical dimension of the nuclear envelope. *Physiol Rev*, **81**(1): 1–19.

35. Dobrovolskaia MA, McNeil SE (2007). Immunological properties of engineered nanomaterials. *Nat Nano*, **2**(8): 469.

36. Jiang X, Rocker C, Hafner M, Brandholt S, Dorlich RM, Nienhaus GU (xxxx) Endo- and exocytosis of zwitterionic quantum dot nanoparticles by live HeLa cells. *ACS Nano*, **4**(11): 6787–6797.

37. Osaki F, Kanamori T, Sando S, Sera T, Aoyama Y (2004). A quantum dot conjugated sugar ball and its cellular uptake. On the size effects of endocytosis in the subviral region. *J Am Chem Soc*, **126**(21): 6520–6521.

38. Jiang W, KimBetty YS, Rutka JT, ChanWarren CW (2008). Nanoparticle-mediated cellular response is size-dependent. *Nat Nano*, **3**(3): 145.

39. Byrne SJ, Corr SA, Rakovich TY, Gun'ko YK, Rakovich YP, Donegan JF, Mitchell S, Volkov Y (2006). Optimisation of the synthesis and modification of CdTe quantum dots for enhanced live cell imaging. *J Mater Chem*, **16**(28): 2896–2902.

40. Nabiev I, Mitchell S, Davies A, Williams Y, Kelleher D, Moore R, Gun'ko YK, Byrne S, Rakovich YP, Donegan JF, Sukhanova A, Conroy J, Cottell D, Gaponik N, Rogach A, Volkov Y (2007). Nonfunctionalized nanocrystals can exploit a cell's active transport machinery delivering them to specific nuclear and cytoplasmic compartments. *Nano Lett*, **7**(11): 3452–3461.

41. Williams Y, Sukhanova A, Nowostawska M, Davies AM, Mitchell S, Oleinikov V, Gun'ko Yi, Nabiev I, Kelleher D, Volkov Y (2009). Probing

cell-type-specific intracellular nanoscale barriers using size-tuned quantum dots. *Small*, **5**(22): 2581–2588.

42. Lovric J, Bazzi HS, Cuie Y, Fortin GR, Winnik FM, Maysinger D (2005). Differences in subcellular distribution and toxicity of green and red emitting CdTe quantum dots. *J Mol Med*, **83**(5): 377–385.
43. Gagnon E, Duclos S, Rondeau C, Chevet E, Cameron PH, Steele-Mortimer O, Paiement J, Bergeron JJM, Desjardins M (2002). Endoplasmic reticulum-mediated phagocytosis is a mechanism of entry into macrophages. *Cell*, **110**(1): 119–131.
44. Dubertret B, Skourides P, Norris DJ, Noireaux V, Brivanlou AH, Libchaber A (2002). *In vivo* imaging of quantum dots encapsulated in phospholipid micelles. *Science*, **298**(5599): 1759–1762.
45. Jaiswal JK, Goldman ER, Mattoussi H, Simon SM (2004). Use of quantum dots for live cell imaging. *Nat Methods*, **1**(1): 73–78.
46. Parak WJ, Gerion D, Pellegrino T, Zanchet D, Micheel C, Williams SC, Boudreau R, Le Gros MA, Larabell CA, Alivisatos AP (2003). Biological applications of colloidal nanocrystals. *Nanotechnology*, **14**(7): R15–R27.
47. Kopito RB, Elbaum M (2007). Reversibility in nucleocytoplasmic transport. *Proc Natl Acad Sci*, **104**(31): 12743–12748.
48. Feldherr CM, Akin D, Cohen RJ (2001) Regulation of functional nuclear pore size in fibroblasts. *J Cell Sci*, **114**(Pt 24): 4621–4627.
49. Terry LJ, Shows EB, Wente SR (2007). Crossing the nuclear envelope: hierarchical regulation of nucleocytoplasmic transport. *Science*, **318**(5855): 1412–1416.
50. Willison JH, Johnston GC (1978). Altered nuclear pore diameters in G1-arrested cells of the yeast *Saccharomyces cerevisiae*. *J Bacteriol*, **136**(1): 318–323.
51. Conroy J, Byrne SJ, Gun'ko YK, Rakovich YP, Donegan JF, Davies A, Kelleher D, Volkov Y (2008). CdTe nanoparticles display tropism to core histones and histone-rich cell organelles. *Small*, **4**(11): 2006–2015.
52. Ishihama Y, Funatsu T (2009). Single molecule tracking of quantum dot-labeled mRNAs in a cell nucleus. *Biochem Biophys Res Commun*, **381**(1): 33–38.
53. Rzigalinski BA, Strobl JS (2009). Cadmium-containing nanoparticles: perspectives on pharmacology and toxicology of quantum dots. *Toxicol Appl Pharmacol*, **238**(3): 280–288.

54. Akerman ME, Chan WC, Laakkonen P, Bhatia SN, Ruoslahti E (2002) Nanocrystal targeting *in vivo*. *Proc Natl Acad Sci USA*, **99**(20): 12617–12621.
55. Evanko D (2006). Bioluminescent quantum dots. *Nat Methods*, **3**(4): 240–241.
56. Cho SJ, Maysinger D, Jain M, Roder B, Hackbarth S, Winnik FM (2007). Long-term exposure to CdTe quantum dots causes functional impairments in live cells. *Langmuir*, **23**(4): 1974–1980.
57. Hardman R (2006). A toxicologic review of quantum dots: toxicity depends on physicochemical and environmental factors. *Environ Health Perspect*, **114**(2): 165–172.
58. Nel A, Xia T, Madler L, Li N (2006). Toxic potential of materials at the nanolevel. *Science*, **311**(5761): 622–627.
59. Comment E (2007). Healthy challenges. *Nat Nanotechnol*, **2**(8): 451.
60. O'Brien P, Haskins JR (2007). *In vitro* cytotoxicity assessment. *Methods Mol Biol*, **356**: 415–425.
61. Jan E, Byrne SJ, Cuddihy M, Davies AM, Volkov Y, Gun'ko YK, Kotov NA (2008) High-content screening as a universal tool for fingerprinting of cytotoxicity of nanoparticles. *ACS Nano*, **2**(5): 928–938.
62. Byrne SJ, Williams Y, Davies A, Corr SA, Rakovich A, Gun'ko YK, Rakovich YP, Donegan JF, Volkov Y (2007). "Jelly dots": synthesis and cytotoxicity studies of CdTe quantum dot-gelatin nanocomposites. *Small*, **3**(7): 1152–1156.
63. Derfus AM, Chan WCW, Bhatia SN (2004). Probing the cytotoxicity of semiconductor quantum dots. *Nano Lett*, **4**(1): 11–18.
64. Chang E, Thekkek N, Yu WW, Colvin VL, Drezek R (2006). Evaluation of quantum dot cytotoxicity based on intracellular uptake. *Small*, **2**(12): 1412–1417.
65. Shavel A, Gaponik N, Eychmuller A (2006). Factors governing the quality of aqueous CdTe nanocrystals: calculations and experiment. *J Phys Chem B*, **110**(39): 19280–19284.
66. Zheng Y, Gao S, Ying JY (2007). Synthesis and cell-imaging applications of glutathione-capped CdTe quantum dots. *Adv Mater*, **19**: 376–380.
67. Liu W, Choi HS, Zimmer JP, Tanaka E, Frangioni JV, Bawendi M (2007). Compact cysteine-coated CdSe(ZnCdS) quantum dots for *in vivo* applications. *J Am Chem Soc*, **129**(47): 14530–14531.
68. Moreau JW, Weber PK, Martin MC, Gilbert B, Hutcheon ID, Banfield JF (2007). Extracellular proteins limit the dispersal of biogenic nanoparticles. *Science*, **316**(5831): 1600–1603.

69. Cedervall T, Lynch I, Lindman S, Berggard T, Thulin E, Nilsson H, Dawson KA, Linse S (2007). Understanding the nanoparticle-protein corona using methods to quantify exchange rates and affinities of proteins for nanoparticles. *Proc Natl Acad Sci USA*, **104**(7): 2050–2055.

70. Qi L, Gao X (2008). Quantum Dot-amphipol nanocomplex for intracellular delivery and real-time imaging of siRNA. *ACS Nano*, **2**(7): 1403–1410.

71. Deshayes S, Heitz A, Morris MC, Charnet P, Divita G, Heitz F (2004). Insight into the mechanism of internalization of the cell-penetrating carrier peptide Pep-1 through conformational analysis. *Biochemistry*, **43**(6): 1449–1457.

72. Gao X, Cui Y, Levenson RM, Chung LW, Nie S (2004). *In vivo* cancer targeting and imaging with semiconductor quantum dots. *Nat Biotechnol*, **22**(8): 969–976.

73. Dubois RN, Abramson SB, Crofford L, Gupta RA, Simon LS, Van De Putte LB, Lipsky PE (1998). Cyclooxygenase in biology and disease. *FASEB J*, **12**(12): 1063–1073.

74. Byrne SJ, le Bon B, Corr SA, Stefanko M, O'Connor C, Gun'ko YK, Rakovich YP, Donegan JF, Williams Y, Volkov Y, Evans P (2007). Synthesis, characterisation, and biological studies of CdTe quantum dot-naproxen conjugates. *Chem Med Chem*, **2**(2): 183–186.

75. Stroh M, Zimmer JP, Duda DG, Levchenko TS, Cohen KS, Brown EB, Scadden DT, Torchilin VP, Bawendi MG, Fukumura D, Jain RK (2005). Quantum dots spectrally distinguish multiple species within the tumor milieu *in vivo*. *Nat Med*, **11**(6): 678–682.

# Index